D0782110

# Heartminded

**Consciously evolving from fear to solidarity**

Dr. Silvia Casabianca
MA, LMCH, RM, MQP, CTP

**Eyes Wide Open, 2020**

**Heartminded: Conscious evolution from fear to solidarity**
**First edition, 2020**

Editor: Tom Wallace - www.tomwallacewrites.com
ISBN – 978-0-359-99891-3
Naples, Florida, USA
Cover image copyright: Sebastian Kaulitzki -123rf.com


EYES WIDE OPEN - Naples, Florida
mail@eyeswideopenc.com
**www.silviacasabianca.com**

***Warnings:***

***Most of the information provided in this book is based on the author's personal opinions and experiences, except when indicated otherwise. Every effort has been made to give proper credit to other author's ideas. Any recommendation offered on these pages has the main purpose of helping the reader make informed decisions.***

***Some of the stories used as examples are real, others are fictional. Only those already published in the media include people's real names. Clinical vignettes are inspired by real stories, but data or features that could lead to identifying the subjects have been modified or deleted, in order to preserve their anonymity and confidentiality.***

***In this book I will not examine the subjects of gender, sexual orientation, or sexual preferences. After all, what I'll be discussing relationships and not who are in them.***

***Also, please be advised that my use of first-person pronouns is random. I don't intend to assign a particular behavior or role to a specific gender.***

## Acknowledgements

I dedicate this book, with gratitude, to each and every person in my personal and my professional life who contributed to my exploration of what love is, how we love, why we can't love as we would like, and what love's possibilities are. I especially thank my clients, who, through all my years of professional practice as a psychotherapist, opened their hearts to me and shared their worries, fears, frustrations, and deep pain. Many of the questions and answers explored in this book come from them. I owe a debt of gratitude also to the teachers in my life—not only my professors but all those who helped open my heart, either by offering me love and understanding or by presenting challenges that fed my curiosity and encouraged me to grow. Without these contributions, I would not be who I am.

Heartfelt thanks to my greatest teacher, my daughter Sandra Silva. In addition to giving me her unwavering love, she helped me research parts of this book. Thanks also to Paulina Díaz, co-creator of our Carpe Diem school project, for her suggestions after reading some pages of this manuscript and because, without her, our educational innovation experiment could not have been carried out.

To Julia Byers, Distinguished Senior Fellow, Global Resilience Institute, Northeastern University and Professor Emerita, Lesley University, and Maria Loffredo Roca PhD, Chair of the Department of Integrated Studies at FGCU, GreenFaith Fellow: both took a time from their very busy schedules to read the manuscript and provide precious feedback that helped improve the book. My generous friend Karen Longordo also provided feedback on the book and used her professional magic on my headshot. To all of them, I am very grateful.

I'm indebted to my editor, Tom Wallace, for his invaluable advice, and assistance, and for his meticulous job on the manuscript.

It'd be impossible to name them all, but thanks to my tutors, mentors, supporters, and mentees for all I've learned from them.

***Books by Silvia Casabianca***

*Regaining Body Wisdom: A multidimensional view*
*Stress Management for Massage Therapists*
*Integrative and Holistic Stress Management for Nurses*
*Health, Illness, Therapeutic Relationships, and Cultural Competence*
*Reiki for kids*

*Si el Arco Falla, la Flecha no Volará*
*El Fin de la Enfermedad*
*Los Cuatro Elementos del Exilio*
*Cerrado el Capítulo*
*Manejo del Estrés para Masajistas*
*Hombres de Papel*

*"Silvia Casabianca is a most compelling author whose multicultural background and education are felt on each page of her book. Her thoughts, ideas, and compassionate concern for the well-being of all are supported by her exploration of seminal thinkers who have informed her life and experience. What is profoundly moving about this book is how Silvia integrates points of view from her Colombian heritage with her Canadian and American perspectives, leading to a global approach to developing empathic skills necessary for our collective world to survive. The author's focus on wellness goes beyond that expressed in Sandro Galia's book, "Well: What We Need to Talk about When We Talk about Health," to the core of what can lead individuals and societies—large and small—to deeply value our humanity and develop a sense of its resilience. HeartMINDED, as the title suggests, is an antidote to the problems of societies, communities, and other groups that have tried to warm themselves around the campfire of technology but have turned their backs on the essentials of compassion, care, and civility. Thank you, Dr. Casabianca, for your extraordinary, creative book that profoundly contributes to the ongoing effort to help people to feel truly well."*

***JULIA GENTLEMAN BYERS EdD, Distinguished Senior Fellow, Global Resilience Institute, Northeastern University.***

*While reading Dr. Casabianca's book, I found myself constantly nodding my head in agreement. She clearly articulates the social ills of our time and seeks to prescribe remedies rooted in the development of compassion and empathy in each of our hearts and minds. She states that, "We need a new, multidimensional paradigm to approach life." This book is a must read for anyone concerned about the future. It is exceptionally well researched and rich in insights and recommendations for how to effect change based in a paradigm of love. To get there, Casabianca provides a deep exploration of love at every level and helps the reader understand the potential power of a new paradigm built on a foundation of love. As she notes, it may be our only hope.*

***MARIA F. LOFFREDO ROCA, PhD, Chair Department of Integrated Studies, FGCU, GreenFaith Fellow***

*We humans seem to be mired in strife. Sometimes, it almost seems a part of who we are as a species. But is that really our natural state? Might it be possible for us to rise above the hatred, greed, and resentments that seem such integral parts of our daily lives? Can we put aside our grievances—in our families, our jobs, and our personal associations—and learn to approach each other with love? Can we come to truly care for each other on a global level? In these pages, Dr. Silvia Casabianca demonstrates that, in fact, humans are hardwired for love and compassion. These gifts reside in our physiology, our chemistry. And they can be nurtured and developed. They can be harnessed and used to solve many of the problems we struggle with—from the interpersonal to the geopolitical. Millennia of human experience have led us to this moment, when we are perhaps finally ready to weave together all we've learned about the mind, body, and soul and come to recognize, embrace, and enact our true, loving nature.*

*Dr. Casabianca's message is one of hope for all. Through the ideas and practices explored in this book, we can heal ourselves, our families, our governments, and even our planet.*

***TOM WALLACE, Editor***

# Contents

# INTRODUCTION

*As for...the fear that compassion will involve you in suffering, counter it with the realization that the sharing of sorrow expands your capacity to share joy as well. When you callously ignore the suffering of others, you lose the capacity to share their happiness, too.*
—Albert Schweitzer

I have fond memories of my grandfather Abel. I have this image of him on one afternoon in particular, looking at me with his dark eyes, his upper lip partially covered by his moustache, standing tall in front of me after a family lunch at his place. When he saw I wanted to talk to him, he sat down and kindly listened to me. I was ten, and I was telling him I wanted to become a writer. His consideration made me feel truly cherished, and although, for reasons that don't matter now, he could not fulfill his promise of teaching me the basics of the literary art, his focused attention helped me to believe in myself.

Growing up, my regard was mostly focused on fascinating men—like my grandfathers, my father, my teachers, the pediatricians who alleviated my pain or vaccinated me to prevent disease, in short, on those men who invited many of my first wonderments. Then, little by little, my life became influenced by women, whom, for different reasons, I found equally or even more admirable. My appreciation for those men stemmed from their brilliance; their knowledge of science, philosophy, or history; and their amusing conversation. Not only did they seem to have answers for any of my impertinent questions, they were frontrunners in the marathon of life, with their intellect, their logic, their ability to be in the world generating bright ideas, managing huge projects, healing the infirm, governing, and erecting bridges and buildings. But my admiration redirected—gradually, quietly—to women who subsequently appeared in my life, women that I found equally or even more fascinating. I especially want to mention the women from the villages and communities settled in the south of the department (equivalent to a state) of Bolivar, Colombia.

In the eighties, I moved from the big city to the countryside with my daughter, who was still a small child, to become a part of a medical team at the well-equipped Medical Center of Specialists, founded by the late doctor Roberto Giraldo, in Magangué, a port on the Rio de la Magdalena. At the center, we supplied quality yet affordable care to people coming from faraway villages where they had no access to health services. Our patients were mostly blue-collar workers, peasants, fishermen, stevedores, lotto ticket hawkers, and hard-working employees who lived in arduous conditions on the banks of the rivers Cauca, Chicagua, and Magdalena, as well as in the tropical jungles on the distant mountains of southern Bolívar.

Also, every other weekend, we traveled up the river in *chalupas* (river motorboats) and then on foot or mules up the mountains, reaching remote areas that lacked the health services that the local, state, or national government should have been providing. Prior to the beginning of our program, the only health services people in these areas had access to were provided by *curanderos* (healers—using mostly folk remedies) and self-made *parteras* (midwives).

In most cases, it was local women who were in charge of organizing our medical brigades, spreading the word about the arrival of the doctors, and taking care of the logistics. They improvised offices and exam rooms under thatched roofs in dwellings that often had dirt floors and no windows. In most cases, instead of an actual door, there was a towel or a cloth covering the doorway into the exam room for privacy. They collected donations from the patients to pay our transportation expenses. They offered us lodging in their own homes and cooked our meals.

These women tended to the fire, cooked in their *fogones de tierra* (stoves made of a mud mortar mix), washed their clothes in the river, took care of both their own and others' children, sewed and patched up clothes. Each of them also took good care of the family animals (usually one dairy cow, a pair of pigs, and a handful of chickens). There was no job that seemed too big to them. They administered injections and played nurse, *sobandera* (bonesetter), healer, or midwife where there existed no other health resources. They became *maestras* (teachers) where there were no schools. Women took positions of authority in those *corregimientos* (small villages) that the men had left to make a living in neighboring Venezuela at a time when the bolivar was a stronger currency than the Colombian peso. They comforted the mourners and recited prayers when someone passed away. These women also gathered people together in times of difficulty, and, on top of it all, they had time and energy to sing, dance, and organize commemorations and festivities. In other, even more remote, wild areas that we only visited every other month or so, we witnessed how women became essential partners to settlers guaranteeing the survival of their men and their families in those remote areas.

I never heard them complaining about their harsh living conditions. On the contrary, I witnessed women who, besides their daily chores, didn't hesitate to work side-by-side with men, sowing and harvesting crops. I met female healers who found their own formulas for antivenom, *pilanderas* who milled the rice with *pilones* (clubs) almost larger than themselves; *lavanderas* (launderers) who carried their laundry in huge *poncheras* (a large bowls) on the tops of their heads and walked barefoot down to the river ravine, disregarding the danger of a treacherous snake attacking them. Their stories, told in the evenings while sitting on stools around the fire, transmitted the wisdom gained through the inevitable struggles in their lives. These women, who worked without pause, provided the community with the fine fabric that held it all together.

It takes a village to raise a child? True. Not only because these women collectively took responsibility for the safety and welfare of the children who frolicked in the streets when they weren't in school but because, in many cases, even when they already had children of their own, without any hesitation they would offer themselves to foster or adopt other children, taking care of them as zealously as they did their own.

But earlier in my life, when I'd become a mother, I'd already taken a peek at those seemingly innate qualities of women. Summoned by my mother, in whose house I stayed the month before and the month after giving birth, I found my world suddenly turned into this sphere of female coprotagonists, coming together to share with me their wisdom about the vicissitudes of parenting and motherhood (my aunts, my mom's friends, our maids). I had never before been the subject of so much care and generosity than during pregnancy and postpartum.

A new and unfamiliar love germinated in my heart the moment my daughter stared at me with those piercing and enormous black eyes that were seeing the world for the first time. Something clicked inside me. With her birth, my visceral way of reacting to life's upsets suddenly decreased in intensity. My girl somehow made me more "paused," increasingly attentive, and at the same time more vulnerable. With her, I began to comprehend what it meant to synchronize with another being: to perceive her hunger, cold or discomfort. The moment she opened her senses to this awesome, complicated, and sometimes perverse world, I felt I'd grown an unconditional—and in a certain way, fierce—heart.

Now I know the new heart I gained with her arrival was not just metaphorical. As I'll show in the following pages, neuroscience explains that the human brain is physiologically equipped to respond to life challenges with an emotional regulation system—the tend-and-befriend system—that guarantees the care and survival of our offspring. This is an advantageous evolutionary device, especially in a time of crisis. And this system is prevalent in women.

Now I know that when I became a mother hormonal changes during labor activated a particular set of neuronal circuits[1] in the brain that encouraged me to tend to my little girl's needs—circuits that steadily increased their synaptic connections in response to her smiles and cuteness and my striving to guess and anticipate her needs. I have no doubt that with the birth of my daughter, new neural networks of empathy and love grew and became more intricate in my brain. I also know that this emotional experience was not exclusively mine, that if evolution led women to develop a stronger inclination to care for others, it's because it represents an adaptive advantage that guarantees the survival of the tribe. (Or in evolutionary genetics terms, the survival of our genes).

I understand how this assertion about women in a polarized era might be misconstrued, and I don't intend to get involved in a feminist argument, responding to claims that women are meant to have a "higher purpose" than being a mother, caring for others, and doing domestic chores. I do believe there is more to women than that, but I also believe an inclination to care for others is our greatest strength.

I also acknowledge that the roles women currently play in the most progressive movements around the globe are no coincidence: we're striving toward the advent of a more just and friendly world in which our children will not only survive but thrive. That's why we care; that's why we get involved. As we'll see later in this book, men tend to have a more active reward (dopamine producing) system, which, in part, explains their motivation to become good providers, to seek positions of power, as well as gratification and pleasure. On the other hand, the neuronal circuits that facilitate connection, empathy, and love tend to be more active in women.

However, I wouldn't dare to assert, as Hillary Clinton did, that "the future is female." This statement poses the risk of alienating the other half of the world. It seems evident that all over the globe, in societies now dominated by corporate greed, women are also taking a lead, demonstrating their proficiency in management and government positions and in traditionally male professions. They are moving forward

[1] Formed by interconnected neurons, networks or neuronal circuits communicate with different brain regions and transmit information (signals) from one region of the body to another.

to play, on a much larger scale, some of the same roles I observed in the women of those remote places in the south of Bolívar, and most of them are driven by their capacity for empathy and compassion for the planet and all sentient beings.

Take, for example, the young Pakistani Malala, who risked her life advocating for the rights of girls to have an education. Or Asnaini Mirzan, in Indonesia, who set out to demonstrate that she could be a mother, a farmer, and a leader who could unite her community. A group of women in the US created the Black Lives Matter[2] movement, calling for racist violence to stop. Mothers have come together in the United States in solidarity to reform gun laws in the hopes of preventing massacres such as the one that took place on Valentine's Day 2018 in Parkland, Florida. Natalia Ponce de León, the victim of an attack that marked her face and life forever, became a human rights activist. Her highest achievement is that, for the first time, Colombia passed a law criminalizing acid attack. And we will talk later about the #MeToo movement against sexual harassment and abuse, also led by women.

Women are coming together around the world to protect water, endangered species, and trees, and to prevent the destruction of the planet. They also advocate for refugees and children. Their ability to respond to crises with solutions growing out of empathy and solidarity makes their contributions, side-by-side with men, invaluable. After all, in order to watch over the planet and guarantee the building of fraternal communities, the two halves of humanity are required.

I'm convinced that if love motivates us, we can certainly transform the world.

Many of the obstacles to the flourishing of a more empathetic, supportive, and loving society have to do with what author Fred Previc calls a "supra dopaminergic society," in which individuals compete in search of pleasure and notoriety, a trend Sigmund Freud would have considered self-destructive. The constant activation of the pleasure-and-reward cerebral circuits, mediated by the production of dopamine, largely explain the frenetic quality of contemporary life, consumerism,

---

[2] The Black Lives Matter movement started on social media right after George Zimmerman, accused of the death of a black teenager, Trayvon Martin, was released from custody. The movement gained strength the following year after the deaths of two black youths, Michael Brown, in Fergusson, MO, and Eric Garner, in New York, both deaths caused by police. The movement also participated in the Baltimore protests in 2015 in the wake of the arrest of Freddie Gray, another young African American, who suffered a spinal injury during the arrest and died soon after. The movement seeks to protect the black population from police abuse.

and the search for novel experiences, as well as the licentious satisfactions, egocentrism, and obsession to obtain success and wealth, leading to the creation of highly competitive environments in which ethics become easily relaxed. The consequences of this greedy and nonstop active lifestyle are the disconnection from our bodies, our neighbors, and other living beings. And it often leads to depression, anxiety, and addictive behaviors.[3]

In the mid-nineteenth century, Charles Darwin had already observed the human capacity to feel what he called *sympathy*. (His use of the term was synonymous with empathy, but that term wasn't coined until later.) Darwin considered sympathy an adaptive advantage that would be extended by natural selection. Current neuroscience confirms that we're biologically equipped for love and compassion and that, because of the neuroplasticity of our brains, we can learn and cultivate these feelings deliberately and consciously.

Clinical observations suggest that children develop empathy concurrently with the acquisition of cognitive, receptive, and expressive language, and fine and gross motor skills. Mastering behaviors sequentially from simple to complex, children advance in adaptive, personal-social, cognitive, and academic areas. In the pages that follow, I will propose that they also learn to love in stages.

However, learning to love requires purpose, discipline, and concentration, as Erich Fromm proposed in his book *The Art of Loving* (Harper, 1956). Human beings could consciously choose paths for emotional regulation (fight-or-flight, competition, or display of empathy), opting for a higher than automatic life, mindfully learning to love, willfully contributing to making the world a better place.

Feeling now at the final stages of my journey, my purpose in writing this book is not simply communicating what I have learned but intriguing others, making them curious by sharing my own questions and inviting a dialogue. Perhaps readers will not find my own queries to be very different from theirs, so I hope that this book will become another opportunity to share learnings, doubts, and discoveries, as happens in the therapeutic relationship.

Educating, writing, and helping others heal are the ways I've expressed my being in the world, always looking to become a kinder individual.

The questions about love probably entered my head when I was a child, as a matter of faith. Still today, I believe that Jesus's admonition, "Love thy neighbor," marks the way to the saving of humanity. My

---

[3] Dopamine plays an important role as a neurotransmitter. But its action and effect don't, as is often erroneously believed, produce well-being or joy. Dopamine is just a molecule transmitting these sensations and emotions.

interest in the subject of love has, of course, evolved. For starters, I feel motivated to explore it because of what I recognize as my shortcomings in the art of loving others. Ambitious as I am, I would like to know everything about it, from biology and biochemistry to cosmology. I want to apply what I've learned not only to my personal life but primarily to the support of my patients and students when they ask for guidance, finding themselves at a crossroads. I would like to fuel their determination to get ahead in their battles. And my feelings about this have to do with my ideas about love, not religion.

With this book I also hope to promote the importance of solidary communities that continuously grow in kindness. I want to suggest attitudes and strategies that could contribute to developing a higher awareness of our role on the planet. The book also includes snippets of the rich, unfathomable information about the neurobiology, genetics, and evolutionary theories of empathy, compassion, love, and the nervous system. Current knowledge supports my claim that, if we humans have evolved toward love and are biologically equipped to experience these feelings, it's feasible to create communities that would advance *Homo sapiens* (wise human being) toward something we might call *Homo amandi* (loving human being).

My strongest tenet is that unless we develop our capacity for empathy and learn to be compassionate, and unless we become capable of loving ourselves and of cooperating with each other, we have no right to call ourselves civilized.

In my practice as a family and couples' therapist, I've often witnessed that wherever patriarchal, power-driven relationships prevail, love faces serious obstacles. I frequently point out to clients how any form of oppression or of emotional, verbal, or physical abuse—which often brings patients to my office—is rooted in a social paradigm that values possessiveness. I see parents who consider children to be their property. I also often see women subjected to relationships in which they lose their power as human beings and become objects, sometimes leading them to defensively express the most negative aspects of their potential. Through domination, a person can force their partner to lose faith in herself, to the point where she might still do things to please him and remain by his side, but out of fear, not love. Force and abuse are never nurturing or loving, and this applies to society in general, not just families.

My professional practice as a medical doctor first and then as a psychotherapist, as a Reiki teacher, and as a practitioner of both the Trager Approach and medical Qi Gong, has helped me build a dynamic, integrative, and holistic vision of our multidimensional bodies but also of love and compassion. A meditation practice also helped me develop the quality of being present. From a humanistic perspective, there may not be a more authentic and profound relationship than the one that can develop between therapist and client, in which a person opens up and exposes their vulnerabilities and wounds to another human being who is, initially at least, a stranger. In the process, they become our "fellow travelers," as Irvin Yalom prefers to call his clients.

Thanks to all of the above, I've been able to learn the importance of being aware of our choices and how we could develop a way to react to life stressors, not from a primary stress response—fight-or-flight—but from a different physiological system that leads us to get along, make alliances, and assist others.

For all the above reasons, I believe empathy can be learned, cultivated, and promoted. If educational institutions want to apply a successful pedagogy, they should establish parenting schools ("parenting circles," we might call them), in which parents, teachers, and therapists would explore ways of raising children without intimidating them. They would also need to promote the inclusion of children in making decisions that affect them and in designing their learning processes. Strategies should be developed to provide children with the opportunity to learn the intrinsic value of cooperating and supporting each other. These strategies would significantly contribute to moving education from authoritarian to democratic models and to solving the problem of bullying.

The ideas shared in this book are, therefore, the product of my personal and professional experience, of much reading and study, and of many conversations and introspections. The book also arises, I suppose, from the audacity to think that I could contribute something original: the hypothesis that conquering the capacity to love is the essence of our journey and that we learn to love through identifiable stages that go along with our psychosocial, cognitive, and linguistic development—if conditions are favorable.

In the process of scrutinizing supporting research for my ideas, I found with joy that I'm by no means alone in this attempt to advocate for the ways of love and compassion, which is also a call to sanity that will guarantee peace, social and ethical progress for humanity, and the subsistence of the planet. On the contrary, empathy, compassion, and love are the focus of much attention in the scientific community today. Many schools around the world are already offering empathy training and compassion-focused therapies.

This makes me hopeful.

At least some of the traditional answers we've given to the classical philosophical questions about our origin, purpose, and path now seem thrown into doubt. I also believe some of the classic existential questions might have been poorly formulated, and our definitions of progress and development need to be reconsidered.

If we're physiologically equipped for empathy—hardwired for loving and cooperating with others—why are we so polarized and lonely? Why have we gotten so disconnected from the planet, from those around us, and from our own bodies? But most importantly, what could we do to remedy the ills *Homo sapiens* endures?

## PART ONE: Our Strongest Instinct

*A coward is incapable of exhibiting love; it is the prerogative of the brave.*
—Mahatma Gandhi

Charles Darwin believed that our capacity to experience empathy (what he called sympathy) was our strongest instinct and predicted that, "This virtue, one of the noblest with which man is endowed, seems to arise incidentally from our sympathies becoming more tender and more widely diffused, until they are extended to all sentient beings."[4]

In his book *Born to Be Good: The Science of a Meaningful Life*, Dr. Dacher Keltner, co-director of the Greater Good Science Center and psychology professor at the University of California at Berkeley, explores the innate power our emotions have to connect us.

In an interview with David DiSalvo[5] about his book, Keltner explained what being born to be good meant to him: "Our mammalian and hominid evolution have crafted a species—us—with remarkable tendencies toward kindness, play, generosity, reverence and self-sacrifice, which are vital to the classic tasks of evolution—survival, gene replication and smooth functioning groups."

These tendencies, says Keltner, are experienced in the realms of compassion, gratitude, fear, shame, and joy and are embodied in our genes and our brain. He expresses hope about the new desire in science to study the human potential for compassion.

Traditionally, natural sciences have focused on the study of phenomena and objects, but a new science is emerging that is more interested in the relationships between elements in nature. The prior science focused on dissecting the subject of study in order to understand its components, which led to reductionism, focusing on the part, not on the whole, on the structure, instead of the function. The new science's operating principles and hypotheses are compatible with network

---

[4] Charles Darwin, *The Descent of Man* (New York: American Home Library Co, 1902).

[5] David DiSalvo, "Forget Survival of the Fittest: It Is Kindness That Counts," *Scientific American*. (September 2009). www.scientificamerican.com/article/kindness-emotions-psychology.

thinking.[6] It is holistic (focuses on the whole) and sees the human being as multidimensional, not isolated, but connected with nature. The old science focused on subduing nature—making it productive for man, no matter the cost. The new science is concerned with making our interaction with nature sustainable. The new science is also interdisciplinary.

Neuroscience is part of this new science that is continuously enriched by the contribution of different disciplines, including anatomy, physiology, biochemistry, pharmacology, pathology of the nervous system, as well as the sciences that explore how our interaction with the environment, our interpersonal relationships, our ancestors, and our history explain our feelings, modify the structure of our nervous system, and define our behavior. Neuroscience can help us understand the purpose, origin, and biological basis of love and compassion, as well as our solidarism.[7]

In this century, remarkable advances in the knowledge of the brain and nervous system are turning neuroscience into a dominant trend within the field of psychology, especially for its therapeutic applications. However, neuroscientists seem to have a tendency to assume that every psychological (mental) process can be explained in terms of anatomy and physiology, even though we're still so far from even understanding—or even agreeing on definitions of—mind and consciousness.

As mentioned earlier, I'm going to use what I'm learning from neuroscience to argue that we are anatomically, physiologically, and biochemically equipped to feel empathy and compassion and love, but I want to warn the reader: it's one thing to explore the correlation between brain activity and mental processes and quite another to say that mind and consciousness are merely products of brain activity.

Hormones and neurotransmitters, such as estrogen, progesterone, oxytocin, dopamine, serotonin, and endorphins, are secreted by the body when we experience feelings of love. Neurobiological processes are involved in erotic love as well as in the creation of attachments between couples, but neurotransmitters and hormones also have a part in other

---

[6] Network thinking: a transdisciplinary model integrating recent advances in psychology, neuroscience, sociology, anthropology, the theory of complex networks, and the newest communication theories. Not an anthropocentric approach, it acknowledges that we're interconnected socially and emotionally with others in relational networks. This integrative model allows us to imagine, reflect, and interact with others and with the world.

[7] A theory that the possibility of founding a social organization upon a solidarity of interests is to be found in the natural interdependence of members of a society (freedictionary.com).

forms of love, such as maternal and fraternal love, even influencing an individual's ability to feel empathy and compassion.

It's also well-established that experiencing (giving and receiving) love and compassion can be beneficial for both our physical and mental health, while several studies suggest that social isolation—in both humans and animals—accounts for one of the greatest mortality risks.[8] It makes sense from an evolutionary point of view: well-being and pleasure resulting from feeling and receiving love contribute to guaranteeing the survival of the individual and the species. Although equipped to feel empathy and express love from birth, and even though history has shown us the most beautiful examples of our human potential for compassion, we face many challenges that prevent us from expressing and nurturing these feelings. More than probably ever before, we're disconnected from our bodies and from others. The inability of individuals to be empathic—when manifested as lack of cooperation and lack of solidarity—is a significant obstacle when our planet's most demanding problems require urgent collective action. This is why it's fundamental to scrutinize what gets in the way of love.

After exploring the reasons that seem to separate us, we also need to examine the opportunities offered by our biological makeup.

Please be advised that I have a radical point of view about our future. I don't see a better remedy to the current state of affairs: either we learn to be empathetic, compassionate, and kind or we destroy ourselves as a species.

## Driven by a need to connect

*Someday, after mastering the winds, the waves, the tides and gravity, we shall harness for God the energies of love, and then, for a second time in the history of the world, man will have discovered fire.*

—Teilhard de Chardin

In spite of the terrible words of hatred we sometimes read in news articles and social media, in spite of how scorching and even destructive political debates can become, and in spite of the multiple wars currently raging and the fact that media are inclined to favor stories of abuse,

---

[8] A. MacBeth and A. Gumley, "Exploring Compassion: A Meta-Analysis of the Association Between Self-Compassion and Psychopathology," *Clinical Psychology Review* (August 2012): 32(6):545-52. doi: 10.1016/j.cpr.2012.06.003.

corruption, and conflict, we still see that we are all searching for love. We're driven by a need to connect, to know we're part of a whole. Even if we're not fully aware of it, as social beings, we instinctively know we need to bond; we want to be loved, to feel needed, and to be useful to others. We want to feel protected, supported, belonging to a tribe.

Several studies conducted with mammals, from small rats to humans, validate the idea that our well-being depends significantly on our social environment and that we suffer when our bonds are threatened or truncated. We now have access to plenty of evidence that we're designed to connect with others.

When we're rejected by a social group, when we become victims of bullying, or when we lose a loved one, we suffer social pain, which demonstrates that connections between humans are not optional or fortuitous but that there is an essential (adaptive) need to create attachments.

In his book *Social: Why Our Brains Are Wired to Connect (Broadway Books, 2014),* UCLA professor Matthew Lieberman says, "Being socially connected is our brain's lifelong passion. It's been baked into our operating system for tens of millions of years." Lieberman is a co-founder of social cognitive neuroscience, a discipline that analyzes how brain function underlies social thinking and social behavior.

Psychologists Roy Baumeister and Mark Leary[9] said the reasons behind our psychological needs for belonging, feeling connected, and forming affective bonds are adaptive demands. They made these interesting observations:

- Once a relationship is established, people are reluctant to break it, even when tension, conflict, or even abuse exist. That is, people prefer to avoid separation, even if there is a high emotional cost to pay.
- When feeling close to others, our thoughts adapt to include aspects of the other in our concept of ourselves until we come to feel that our destinies are intertwined.
- When we don't have close relationships with others, we suffer.
- Separations, even if brief, produce discomfort and sadness.
- Relationships carry significant emotional weight: we are happy when things are going well in relationships; we tend to feel miserable, anxious, jealous, when there is conflict.

---

[9] R. F. Baumeister and M. R. Leary, "The Need to Belong: Desire for Interpersonal Attachments as a Fundamental Human Motivation," *Psychological Bulletin* 117(3) (1995): 497–529. Retrieved July 25, 2019: https://pdfs.semanticscholar.org.

- Statistics show that those who maintain a relationship are healthier, less stressed, and have a longer life expectancy.
- People prefer to have few, but very close friends and a greater number of acquaintances, deeming quality to be more important than quantity. This is because establishing a bond takes time and requires an investment of effort and energy.
- When a relationship breaks down, people tend to look for a new one.

Baumeister and Leary concluded that human beings are motivated by a need to belong, a strong desire for forming and keeping long-lasting interpersonal bonds. They cited attachment theories that have extensively studied how babies spontaneously create bonds with their caregivers. This need was first studied and described by the psychiatrist Sir John Bowlby, who opened the doors to a deeper understanding of the fact that we are social animals and that the first few years of a child's life might determine their future mental health.

Studying children who had been separated from their parents during World War II, Bowlby found that those children raised in orphanages had cognitive delays, problems regulating emotions, and difficulty relating to other people. In a monograph for the World Health Organization entitled *Maternal Care and Mental Health* (1951), Bowlby proposed that a young child needed the close and constant presence of their mother (or substitute), in which both would find satisfaction and joy.

In the fifties and sixties, in his Wisconsin laboratory, Harry Harlow[10] explored the nature of love, trying to understand how bonds between infants and their mothers were formed. He first observed how monkeys that were kept in total isolation developed social deficits. Then he designed inanimate surrogate "moms" made of either wire and wood or foam and soft terrycloth. Newborn rhesus monkeys were fed through these surrogates. After his experiments, Harlow concluded that the infant's physical contact even with a terrycloth or a wire mother, was as or more important to its well-being and development than the nutrition it received. His work supports the idea that the relationship with the mother is more emotional than physiological and is related to the care the child receives. Harlow also observed that it was difficult to compensate for any early loss of emotional security.

Contemporary authors and researchers Daniel Siegel and Helen Fisher are at the forefront of the studies on attachment (see recommended books). Siegel has emphasized that children who develop a secure bond

---

[10] *"Harlow's Classic Studies Revealed the Importance of Maternal Contact.*" Retrieved July 21, 2019: www.psychologicalscience.org.

with their parents know they can come to them when support is needed. This enables them to later empathize with others.

In *The Whole-Brain Child: 12 Revolutionary Strategies to Nurture Your Child's Developing Mind, Survive Everyday Parenting Struggles, and Help Your Family Thrive* (Delacorte Press, 2012), Siegel explains, "As children develop, their brains 'mirror' their parent's brain. In other words, the parent's own growth and development, or lack of those, impact the child's brain. As parents become more aware and emotionally healthy, their children reap the rewards and move toward health as well."

We know that most hunter-gatherer societies focused on behaviors that benefitted their whole communities, not individual interests. This inclination is still seen in indigenous communities in much of the world.

British social epidemiologist Richard Wilkinson[11] says that, "in more equal societies, where there's a stronger community life, less violence, and more trust, people give higher priority to the common good."

As we have an innate need to connect with others, to feel part of the group, shame and embarrassment tend to ensure conformity to society's rules. Many aboriginal people—in particular those who still perceive society as a community—tend to rely on effective socialization and self-regulation. Some tribes air the individual's transgressions in public, aiming at correcting behaviors that are not beneficial to or that go against the community as a whole.

In a study by researchers from the University of Illinois, Ed Diener and Martin Seligman[12] found that among 222 college students, the self-reported "very happy" people were highly social and had stronger romantic and other social relationships than less happy groups. They were more extroverted and less neurotic. These and other studies seem to confirm that the degree of life satisfaction we experience correlates with our capacity to connect to other people. We are social beings.

Psychoanalyst Erich Fromm probably got it right in *The Art of Loving* when he said that our separation from nature threw us into a permanent state of uncertainty. We are constantly seeking to overcome a feeling of separation, and a failure to bond with others could drive us insane.

Although we long to connect with others, we have more difficulty doing so nowadays than we did in times past. We feel more and more separated today, coexisting on a planet from which we also perceive ourselves as disconnected.

---

[11] See Brooke Harvis's interview with Richard Wilkinson, "Why Everyone Suffers in Unequal Societies," *Yes! Magazine* (March 2010).

[12] E. Diener and M. Seligman, "Very Happy People." Psychological Science, (Jan. 1, 2002).

Charles Eisenstein (charleseisestein.net) a public speaker and gift economy advocate, has suggested we live in a world in which psychology considers us a mind inhabiting in a body; religions preach that we are souls incarnated; physics, that we are matter determined by impersonal forces; biology, that we are like robots made of flesh and bone, programmed by genes for the benefit of a reproductive interest; and economy, that we are rational players seeking to maximize our own financial interests.

However, the new science has begun to prove that we can overcome the perception of separation, which hurts and greatly contributes to many of the ills afflicting humanity. We see ourselves separated by gender, national borders, origin, beliefs, skin color, and social stratum. And the modern devices with which we try to bridge the gaps (Facebook, Instagram, and Twitter, for example) are certainly insufficient for that task, if not counterproductive.

## Our disconnection

We are spending most of our lives in a state of disconnection. We're disconnected from our sensations, perceptions, impulses, emotions, and even our thoughts. We're disconnected from what we say, from those around us, from the planet, and from our bodies.

Love requires an openness to life and to others, but living in a state of defensiveness and mistrust, constantly competing with others, is contrary to that openness.

How often we find ourselves mired in worries, lost in our thoughts, obsessed with the past, immersed in our plans for the future, driven by our desires and fears, and confused by our need to have fun. We live mindlessly, acting on automatic pilot.

It's no secret that the world is becoming increasingly polarized. Our awareness of separation grows each day. The spending habits we have adopted disregard the responsibility we have to care for the planet.

### Neglect of the planet

Our fall from paradise—as a metaphor—seems to refer to the time when we began to separate from nature, which seemingly happened when humans transitioned from hunter-gatherers to shepherds (Abel) and farmers (Cain) during the Neolithic Revolution. Adam and Eve hadn't bitten the fruit of the tree of the knowledge of good and evil, but that of the tree of discord. The progress of agriculture, the concurrent rise of class societies and private property, the creation of states and armies, and the recognition of new types of relationships between men and women soon followed. The family was restructured to serve the interests of a rising class wanting to maintain control over their property.

In a modern world where the population has been progressively displaced toward urban life,[13] we've not only lost the acuity of our senses, but also much of our instinct and intuition. Let's take the example of a hunter: he must learn to listen to the animal that stealthily approaches; he must recognize the marks it has left on the ground, refine his ear to identify the location of a sound that may indicate the proximity of danger or the nearness of the prey. He needs to be able to see and feel the signs his target has left on its trail. As a stalker, he needs a visual sharpness that allows him to detect a target at a great distance and pursue it with his eyes. A refinement of his intuition, his senses, his abilities, makes him a more efficient hunter.

But this was true of hunters who lived in the open most of the time. Modern man dwells inside boxes—different size and shapes, but still boxes—that isolate him from his surroundings. Most of us "hibernate" in artificial climates, and most of the time, we look at nature through polarized glasses. We no longer know what the true colors of life are. We've lost our capacity to adapt to the outdoors. As a result, both instinct and intuition are gone. Our senses have deteriorated. We rely on external gadgets to make up for the loss of our natural wisdom.

Our disconnection from nature in modern life is such that we are unable to anticipate the impact that material progress has on our lifestyle, health, and the planet from which we derive our sustenance. We can't predict the effect a hamburger or, in general, a diet rich in lipids, could have on our immune system or arteries, even though science tells us that such a diet increases the risk of a heart attack or an embolism. We disregard science in the pursuit of immediate gratification.

The news tells us about climate change, melting glaciers, increasing temperatures of the oceans. In the summer of 2017, the largest iceberg in the world—in Antarctica—split up, and in 2018 the Northeastern United States was hit by heavy storms of ice, rain, and snow, apparently caused by Arctic warming. (That year it began to melt early, in February.) We know that sea levels are rising and the coastal cities on the Gulf of

---

[13] According to the United Nations' department of economic and social affairs, the urban population in 1960 was a 34 percent, while in 2014 constituted 54 percent of the global population, and it continues to grow.

Mexico and the islands in the north of the Caribbean have been affected by the most intense hurricanes and tornadoes ever. The most recent, the catastrophic category five hurricane Dorian (September 2019) that left about 70,000 people homeless in Grand Bahama. We're witnesses to more earthquakes, and ravaging fires (related to droughts, deforestation, and deregulation of environmental protections). Endangered species, and toxic algae blooms ("red tide" in my region, caused by algae called *Karenia brevis*), have all become phenomena of unprecedented intensity.[14]

The key question has been whether or not these occurrences are a consequence of human activity.[15] Most evidence points to an affirmative answer, but some governments have chosen not to see this.

One of many examples of the disconnection between our lifestyle and the impact we have on the planet is the comfort with which we buy products packed in plastic because (shrug) we can throw the containers in recycling bins. But where is this waste going to end? Much of the plastic that we throw away is transported (with a high fuel expense) to recycling centers. In 2018, the *New York Times* reported that "hundreds of local recycling programs in American cities and towns are collapsing." The recycling process itself consumes energy, and the cost is now being transferred to the consumer. Municipalities are learning how much more expensive it's to collect, sort, and process recyclable materials than to just send them to landfills. In many cases the plastic ends up being shipped all around the world from, Turkey to Malaysia, in huge tankers that leave a trail of lethal oil in the water, to be deposited on the fields of New Zealand or Third World countries.[16]

In January 2018, China banned the imports of foreign waste (China was the global hub for recyclables but complained that waste received was not properly sorted out and plastic contained cheap, low-end

---

[14] In just a month (August–September 2017) three hurricanes, Harvey in Texas, Irma in Florida, and María in Puerto Rico, caused immense damages (calculated at about 500 billion dollars). Damages caused by Maria are considered the worst disaster ever registered in Dominica. At the same time, a very dry summer ended with fires that affected 10,000 buildings and houses and 47,000 acres in 2017. New fires are ongoing.

[15] "The footprints that humans have left on Earth climate are turning up in a diverse range of records and can be seen in the ocean, in the atmosphere, and on the Earth's surface." Union of Concerned Scientists.

[16] I hesitate to use the term Third World, coined in the fifties, because its meaning has changed. However, I'll use it to designate a group of countries that are less developed technologically and where the living conditions, health indicators, and mean income of most of the population is the least favorable.

materials). The US, Europe, and Japan are having trouble finding alternative destinations. The European Union is considering a tax on plastics, and some countries have started to ban the use of plastic bags, cups, plates, straws, and bottles. To give you an idea of the dimension of the problem, by 2016, it was estimated that around 4.73 billion plastic cups were thrown away every year in France alone.

Our use of plastic is affecting animals. For example, the Pacific Ocean draws approximately ten metric tons of plastic fragments to the beaches of Southern California. Birds, turtles, seals, and other marine animals, deceived by its smell and appearance, mistake plastic debris for food, and a significant number of animals die from malnutrition, chemicals in plastic, or intestinal obstruction. In other cases, animals get stuck or entangled in plastic objects such as fishing nets. At present, these issues are affecting about 115 marine species.

The lack of regulation of certain industrial processes (production, waste disposal) is also responsible for both pollution and the consequences of the presence of plastic in the environment.

Nearly two hundred nations signed a United Nations resolution in Nairobi (2017), to eliminate plastic in the oceans. They estimate the task will take at least thirty years, but this may not be fast enough to avoid environmental disaster. Presently, countries like Spain don't know what to do to dispose of the millions of plastic bottles thrown away by residents and visitors every day. By the time you read this, it's very likely the numbers will be worse. However, markets are still filled with shampoos, alcohol, medicines, supplements, and other products in plastic containers that we seem to have no choice but to buy and take home.

There is consensus in the scientific community (expressed by the Intergovernmental Panel on Climate Change, the IPCC) that human activity is indeed modifying the environment and damaging the planet. However, in the US, legislators, the forty-fifth president, and many in the media still question these conclusions. Numerous actions taken or proposed by the Trump administration are rolling back Obama-era policies aimed at either slowing down climate change or limiting environmental pollution. Federal funding for science and the environment is also shrinking. The interests of large corporations, seeking to avoid the costs of the reengineering necessary to prevent future emissions of greenhouse gases, are behind this problem.

A careless appropriation and abuse of resources and mindless consumerism ignore the impact we're having not only on the planet but our future quality of life.

In the major cities of countries like the United States, up to 40 percent of the food produced is dumped (thirty-one million tons of food added to landfills each year). The easy accessibility of tap water leads us to ignore the consequences of our insensible waste of this resource. For example, a faucet leaking ten drops every minute wastes 526 gallons of water per year. An expanding urban population is making these problems worse.

## See no stranger

*The main point about civility is...the ability to interact with strangers without holding their strangeness against them and without pressing them to surrender it or to renounce some or all the traits that have made them strangers in the first place.*

—Zygmunt Bauman

The media constantly inform us about acts of terrorism, wars, people displaced by violence, refugees, famines, natural calamities, human and drug trafficking, mass lay-offs, corporations that sink overnight or merge to form larger and frighteningly powerful entities. All of these are symptoms and consequences of our disconnection as humanity.

In January 2018, the then prime minister of the United Kingdom, Teresa May, created a new position, a Ministry of Loneliness. More than nine million people in the UK suffer, either occasionally or permanently, from loneliness, according to a report published by the Jo Cox Commission on Loneliness. And loneliness seems, more than anything, the product of our inability to connect with others.

We're not only isolating ourselves; we're regrouping.

Armed with recent demographics, journalist Bill Bishop published *The Big Sort: Why the Clustering of Like-Minded America Is Tearing Us Apart* (First Mariner Books, 2009). When he looked at the electoral results of the last thirty years, he observed that Americans have grouped by class, skin color, and beliefs in increasingly homogeneous communities. This has happened not only at the region or state level, but by city and even neighborhood. His data has been confirmed by other reporters, such as Corey Lang and Shanna Pearson-Merkowitz of the UK,[17] who also predict that this tendency toward segregation will be generalized along party lines. People are choosing neighborhoods (and churches and news programs) that are compatible with their lifestyles and beliefs. This type of grouping prevents the new generations from being exposed to different opinions and views of the world. The phenomenon is

---

[17] The LSE US Centre's daily blog on American Politics and Policy https://blogs.lse.ac.uk/usappblog/.

happening throughout the nation. For example, in rural West Texas, a fifty-acre community development (Paulsville) was created in 2008 to provide homes exclusively for followers of Ron Paul, libertarian presidential candidate at the time.

Bill Bishop suggests that the outcome of this trend has been a notoriously dangerous polarization of the population, a decline in tolerance, and an increase in extremism.

Facebook designs algorithms that select what's displayed in my wall and shown to my friends and family. I will usually get to see more posts from people who think like me and fewer from those who have different opinions, or more personal comments and pictures, than political news. The result is my followers and the people who like or click on my posts back my opinions, but my posts rarely reach those who think differently. Several countries have adopted what's known as a balkanization of social media or splinternet—meant to block, filter, or redirect certain topics—which causes us to live in separate microcosms, with narrower visions. It denies us the opportunity of enriching ourselves with differing ways of seeing the world and it has become a political instrument to perpetuate power in the hands of a few.

In a sociopolitical climate of constant change, corruption in the highest spheres, mutual distrust, and unrelenting competition, we feel easily judged, criticized, and excluded. This also constitutes an obstacle when trying to connect with others. I might distrust others because I suspect they want what I have (my money, my partner, my position at work, my influence). Since others have abused me in the past, betrayed me, abandoned me, rejected me, I can't expect otherwise. In the midst of this mistrust, I keep my guard up. I don't show my vulnerability. I might choose not to connect with those different from me but to group with like-minded people. The paradox is that vulnerability actually connects us, humanizes us. We need to change the paradigm that the intellect is what makes us strong. True strength doesn't come from our physical bodies and brains, which inevitably deteriorate, but from experiences and feelings that eventually make us capable of empathy and prompt our indignation about social injustice or ignorance and ambition. We all have a soft side, and that's just fine. We are yin; we are yang.

## The culture of the envelope

Not only have we progressively disconnected from each other and the planet, we've also stopped listening to our bodies. We've forgotten how to lead a rhythmic life. We don't eat when we're hungry but when the food is available or when it's noon. We don't sleep when tired; that's what caffeine is for. We turn off symptoms with medication instead of trying to understand their root cause. We've lost body wisdom. Our gardens are more for adornment than for receiving our daily dose of sun or for planting trees that purify our air. Instead of exposing our skin to the sun, which would transform skin tocopherols into vitamin D, we take a supplement. Instead of drinking orange juice, we look for vitamin C capsules. If something upsets our stomach, we just take an antacid or digestive enzyme instead of eliminating from our diets the foods causing problems.

We've also stopped trusting the wisdom of the body, no longer listening to its inner healer. We believe that our doctor is the expert in our own body, and we allow specialists to manage our health. I often see patients unable to decide on a course of action because what reason prescribes goes against what their heart, their instinct, or their dreams shout. Society (which strongly echoes the parental voices lodged in your mind) sometimes prevents you from seeing the red flags or what's best for you. We end up not doing what our hearts and souls really need and want.

I believe the above three types of disconnection (from the body, from others, and the planet) are interrelated and lead to deficiencies in our ability to nurture ourselves, love our neighbors, and protect and preserve the environment.

*Development* (a misnomer) has given rise to the adoption of new values, which have a clear detrimental impact on the evolution of the individual and the culture and are very different from the knowledge of our ancestors, who recognized the need to preserve, honor, and care for the planet. But we still call ourselves civilized. I'm using the word *culture* to refer to the ways in which communities explain and make sense of their experiences. Culture, as defined by English anthropologist Edward B. Tylor in his book *Primitive Culture* (John Murray, 1871) is "that complex whole which includes knowledge, belief, art, law, morals, custom, and any other capabilities and habits acquired by man as a member of society."

This disconnection we have created is based on an illusion. In 1973, in his essay "The Cosmic Connection: An Extraterrestrial Perspective," the astrophysicist and author of *Cosmos*, Carl Sagan said, "Our sun is a second- or third-generation star. All the rocky and metallic material we stand on, the iron in our blood, the calcium in our teeth, the carbon in our genes, were produced billions of years ago in the interiors of a red giant star. We are made of star-stuff."

The still-predominant reductionist paradigm feeds the perception of separation from our surroundings, including other people, and convinces us that we're merely individual beings, divided, segments. We've fooled ourselves into denying we're all stardust and that what I do to you I'm doing to myself too, that what I do to the planet affects me.

Fortunately, we're coming to understand that reductionist science, which until recently we thought irrefutable, is questionable and that we benefit by adopting more holistic (*holon* means complete, total), systemic, and comprehensive perspectives. A multidimensional and holistic perspective of health, disease, education, politics, and our relationships with others and with the world goes beyond what current science could even presently explain (measure or corroborate).

Adopting a holistic approach can transform our relationship with our bodies and the environment. If we were more open to ancient cultures, we'd see that Buddhists, Taoists, and Hindus offer us invaluable pearls of wisdom, treasured for generations, and with a universalist perspective. They teach us, for example, that a frugal, moderate discipline and lifestyle, a conscious existence (monitoring our minds), can keep us physically, emotionally, and mentally healthy and is good for the planet. The four components of love in Buddhism are joy, compassion, equanimity, and benevolence, which allow us to connect with others and the environment from a kinder heart.

The Bible also preaches frugality, which some people may interpret as paying less for stuff. The real meaning is having less (only what's necessary), avoiding waste, and not allowing our happiness to depend on what we own. This is also good for the planet.

In psychology and social sciences, we're also approaching a more down-to-earth vision of love and relationships, an understanding that individuals can connect with others without exposing themselves to be hurt.

I will be able to take responsibility for my feelings and experience joy in relationships as long as I can fully express my essence and be myself in the presence of another. It makes all the difference in the world if I learn that it's healthier to choose a companion, friend, neighbor, colleague, or family member who won't judge me and in front of whom I don't need to hide my feelings or thoughts or appearance in order to be loved. In other words, if I learn to be with people capable of accepting me as I am, who love me because of who I am. And if I make a mistake and choose a wrong pal, if someone mistreats me and becomes a toxic presence in my life, it's also important to know that the fairest and healthiest thing to do is to get away.

It might not be necessary to know why or when our instinct and intuition got clouded, or when or why our human relationships became utilitarian, or how we came to have a minimal or neglectful relationship with nature. But it's crucial to overcome this rift between us and our bodies, between us and our neighbors, between us and our planet. It's critical to regaining the natural wisdom through which we keep our inner healer attuned.

How could love and solidarity prosper in a competitive and polarized world where it's become so difficult to bridge the gap between us and those who don't think, live, feel, or vote like me?

## The lie of pretend intimacy

A while ago, I attended an event with indigenous leaders, an opportunity to come together to honor tribal wisdom and increase awareness of our connection to the earth. After the event concluded for the day, a lady I'd met there walked to the bar proclaiming her need for a daily cup of wine—a popular misinterpretation of studies that, a few years ago, recommended not exceeding a daily glass. She told me very happily, "Alcohol is the best way to make friends."

I don't drink, and I know I might have frowned when I heard her. I didn't say anything, but I thought, "The social connection you make with people who drink with you is not a friendship. It's a mirage caused by the effects of alcohol. Maybe you get the illusion of closeness, close contact, and even a sense of belonging to a group, but you can't call it friendship."

What a patient who had been sober for a year told me shows that we can overcome that illusion. He said, "Since I've been sober, I've realized my coworkers never were real friends, and what's worse, it's taking a lot of work to relearn how to socialize while being sober. Now I want to be around other people. I feel out of place among those who are drinking, but I also feel odd hanging out with the people from Alcoholics Anonymous. I even realized that while I was a drunk, I maintained unhealthy relationships because I was trying to conquer my unresolved fears of intimacy by drinking alcohol."

Some of those who go to bars on the weekend often turn to alcohol with the hope (conscious or unconscious) of getting away from their inhibitions and figuratively extending their hand to connect with others (sometimes getting into trouble not so figuratively). Notice that it's not uncommon for the drunk to suffer from what in Spanish is called *querendona* (being annoying, clingy, flirty) when the drinks begin to take effect. Alcohol usually helps to stifle the fear of approaching other people. Unfortunately, a few drinks more and the effect is exaggerated: a feeling of warm proximity is disrupted when the others flee from the daring and tiresome *borrachito* (drunk).

A side note: different to what happens with other mammals, alcohol might have both inhibitory and disinhibitory effects on aggression, social conduct, and impulsivity in humans. But alcohol is not enough to explain the behavior displayed by a person under its influence. There are cultural, personality, and family factors that influence the effects of alcohol (and other drugs).

Under the influence of certain substances or in some mental states, fear can be temporarily overcome, and the notion of personal space tends to vanish. The sacred barrier, behind which we feel adequately equipped to face the world, is crossed. The risk is feeding the fear of getting closer because, under the influence of intoxicating substances, the proximity to others might become a dicey proposition: maybe you end up in bed with the one you least wanted, perhaps someone could hurt you. How many marriages and careers have been destroyed by the disinhibition leading to transient infidelities because an aroused body sought forbidden closeness while drowning and quieting the conscience?

Instead of leading to connecting with others, the consumption of substances may achieve the exact opposite. We might dare to speak truths in a brutal way, get into a stupid fistfight with a friend, end up in jail, or, worse, cause a fatal accident after getting behind the wheel while drunk. And all this can happen when all we were doing was looking for a little closeness, a little fun. Using alcohol and drugs to get closer to others turns out to be a failed strategy. And yet, we use alcohol everywhere: while eating dinner with friends, in the celebration of a birthday, on a romantic date, and even at fundraising galas aiming to help victims of domestic violence (which itself might be facilitated by the disinhibition of aggression caused by alcohol).

With urbanization, drinking alcohol has also become widespread among indigenous populations. This happens throughout the Americas and beyond, and it has played an important role in the disintegration of indigenous communities. On one occasion I heard a group of Eskimo grandmothers lamenting about what civilization had meant to their people. "Our traditions are lost," said one. "People have lost respect and dignity. Men have become lazy to work and prefer to be drinking and smuggling alcohol and cigarettes."

In her lecture "The Future of Modern Love" (offered at the Psychotherapy Networker, 2018), Belgian psychotherapist Esther Perel mentions that while urban life has brought us unprecedented individual freedom, it's also responsible for our isolation, our distrust of others, and our segregation as human beings. She suggests that the loss of our sense of belonging to a community largely explains the quality of modern relationships, ones that no longer follow the traditions or social conventions typical of rural communities. The terms of the relationships have become negotiable. For the majority, marriages ceased to be an economic enterprise and became a romantic initiative, which means progress, Perel says. However, the isolation typical of urban life places enormous expectations on the couple. As much of the social capital is lost (because the tribe is dissolved), the couple tends to become "everything and all" for each other. One must now provide the other the emotional and physical resources that the village and extended family formerly provided. If intimacy used to be the result of coexistence, now the other must also become the resource that supplies all of my connection needs. My mate must make me feel that I'm worthwhile and that I count. They must also be the remedy for my existential solitude. This new and absorbing romantic love is a recipe for disaster, predicts the author. Expectations are impossible to meet. Rejections and breakups are much more painful.

## The agonizing tribe in the global village

Both anthropology and sociology focus on contemporary indigenous communities, looking for indicators of the uses and customs that probably existed before humanity evolved toward stratified societies (slavery, feudalism, capitalism). Archaeologists study remains and artifacts preserved through time, trying to understand what changes in social structures and organizations have occurred throughout history. Now we know that, about ten thousand years ago, favorable conditions existed for nomadic populations to settle in the area of the ancient Middle East. When they started to live in more permanent settlements, these communities learned to domesticate animals and grow food; they invented agriculture. Anciently, social and political structures existed that favored community or combined labor, and the distribution of the product of work was also communal. Human beings were hunters, fishers, or gatherers, working together. No private property existed. Advances in agriculture opened the possibility for farmers to manage land and create food surpluses.

The evolution of the concept of ownership and the desire to secure one's property eventually gave way to the organization of states and armies. The advancement of agriculture and the social changes it brought gave way to the stratification of societies and the appearance of powerful leaders who waged war. Very few foraging communities still exist.

When we take a look at the cosmogonies and worldviews held by surviving indigenous tribes, we commonly run into ideas and values that are very different from those current in most modern Western societies. The Kogi, for example, who consider themselves our elder brothers, predict that our lifestyle will end life on the planet. The Kogi are descendants of the Tayrona culture and have inhabited the Sierra Nevada de Santa Marta mountain range in Northern Colombia since before the Spanish conquest. They have managed to preserve their traditional ways of life. Their ecological and political activism has succeeded in protecting the Sierra from "development."

A *mamo*, a spiritual leader, a kind of priest, is considered an intermediary between celestial (or nature) forces and human beings. His role is to keep the tribe united and to prevent the disintegration of the cosmic order. His authority is based on his wisdom and his ability to maintain the harmony and connection of the people.

Most indigenous societies still preserve their traditions and customs, through which their communities maintain a more harmonious relationship with nature than we do. The bands of hunter-gatherers form a type of society characterized by companionship. Food, residence, company, and memories are shared without compromising the autonomy of the individual. Egalitarianism predominates.

Other tribes barely preserve a sense of community, some traditions, a dialect. Those that have joined modern civilization suffer painful disintegration processes. For example, of the 364 linguistic variants existing in Mexico, "64 are at a very high risk of disappearance, while 43 of them are at high risk, that is to say that one third of the linguistic variants spoken in the country could disappear in the coming years," according to Eva Caccavari Garza's article "Lenguas yumanas: crisis de la diversidad lingüística en Baja California" (Yumana languages: a linguistic diversity crisis in Baja California), published by the digital magazine of the Universidad Nacional Autónoma de México on February 1st, 2014 —UNAM (National Autonomous University of Mexico). Indigenous communities affected by economic globalization, socioeconomic marginalization and bullying in the schools are losing their dialects.

It's moving to see that the tribe, both in its literal and its metaphorical meaning, agonizes in a world where today you are here, and tomorrow you might be speaking from anywhere in the world. The so-called developed countries have given rise to a different kind of personality with a clear impact on the development of the individual and the culture. Canadian philosopher Marshall McLuhan coined the term "global village" to refer to how the electronic flow of information through digital media facilitates increasing interconnectivity on a global scale. However, even though we now have access to information we previously didn't have access to, living in this global village has its own perils. On the one hand, it allows us to see other cultures' peculiar ways of thinking and existing, offering us a unique opportunity to connecting with people in any latitude. On the other hand, it contributes to the creation of fears and the spreading of mistrust. If globalization tends to homogenize cultures and beliefs across nations, at the same time it makes our differences more tangible, which stresses a hierarchy of knowledge. This global village deceives you with the illusion of connection through the use of social media, while at the same time it tends to further a painful disconnection in personal relationships.

But, as discussed before, that disconnection seems to have begun long ago, when man was banished from paradise (where he lived in harmony with all other living beings) and it progressed as nomads settled and eventually began to discover the written word, humans created urban centers, and developed agriculture. A more sedentary life led to the numbing of instincts and the obscuring of intuition.

I wonder if the original purpose of all major religions was the transmission of the history and teachings needed to bind collectivities together, to set norms that would replace, if not the lost instinct, then that inner wisdom that came from an intimate connection to the natural world. When the oral tradition was exclusive, the elderly were respected because they carried in their memory the experience and wisdom of the group.

The origin of the word *religion* has been disputed, but I'm inclined to accept the notion that it comes from the Latin *re* (return to) and *ligare* (connect). In principle, religions promote a morality based on a sense of connection with the whole (call it god, spirit, immaterial world, universe) and also a connection among its followers or congregation. It would seem that over time, the written word supplants the sage's voice, and the scriptures convey traditions, histories, and norms that, thanks to the patriarchs' hard work, were passed down from generation to generation. It may have had the purpose of keeping the memory of the origins and sustaining a certain social order through the transmission of norms. That's difficult to tell, but religions are certainly focused on teaching people how to live, how to relate harmonically, how to maintain health.

It's interesting to note that the admonition to love one's neighbor is not limited to the New Testament, it's also in the scriptures of various major Abrahamic religions (Christianism, Judaism and Islam). "Take no revenge and cherish no grudge against your own people. You shall love your neighbor as yourself," says the Bible (Leviticus 19:18). "When an alien resides with you in your land, do not mistreat such a one. You shall treat the alien who resides with you no differently than the natives born among you; you shall love the alien as yourself; for you too were once aliens in the land of Egypt." (Leviticus 19:33–34).[18]

Similar messages appear in the Old Testament:

Proverbs 14:21: "Whoever despises the hungry comes up short, but happy the one who is kind to the poor!"

Proverbs 3:29: "Do not plot evil against your neighbors, when they live at peace with you."

The Torah and the Qur'an also call you to be a good neighbor:

"Love thy neighbor as thyself—only if he is your neighbor—that is, if he is virtuous but not if he is wicked—as it is written, the fear of the Lord is to hate evil" (Torah, Prov. 8:13).

"And to parents do good, and to relatives, orphans, the needy, the near neighbor, the neighbor farther away, the companion at your side, the traveler, and those whom your right hands possess" (Qur'an 4:36).

The themes of loving kindness and compassion have been present in the Buddhist teachings through the generations, especially in the Mahayana tradition, in which the main goal of the aspirant is the enlightenment of all beings. Hinduism preaches that selfishness should be renounced and love should be practiced without expecting anything in return.

Our Western society has developed a high appreciation for academic, intellectual knowledge, even if the topics might not have practical application for the learner. We prefer it over knowledge derived from experience, over wisdom transmitted orally or through deeds.

Pope Francis said, "May it never happen again that the religions, because of the conduct of some of their followers, convey a distorted message, out of tune with that of mercy." But, unfortunately, throughout history, the scriptures' authority has often been used in ways that subvert their initial purpose and lead to religion-inspired politics. Some of the most devastating terrorist attacks in recent years were perpetrated in the name of God or in the name of Allah, for example.

Although many religious leaders seek to unite with their teachings, other doctrines, when applied dogmatically, segregate us.

How very often these days doors are slammed in the faces of our neighbors, as has recently happened at the US-Mexican border to asylum seekers. Many are being forced to wait in Mexico, where they are not always safe and thousands of children among these people have been either separated from their parents or placed in nightmarish shelters. Unacceptable!

---

[18] All quotes from the Bible come from the New American Bible revised edition (NABRE).

## Separation: an illusion

*Our separation from each other is an optical illusion of consciousness.*

—Albert Einstein

Each one of us, in their own way, pursues the light (the truth, transcendence, god, source, or love) represented in the center of this drawing, at which the lines converge. These converging rays represent our individual paths. The closer we come to that source in the center, the smaller the distance we perceive between our paths, and the lesser the illusion of separation between us. On the other hand, the farther we are from the center, the more significant that perception of separation becomes.

Professor and medical Qi Gong doctor Jerry Alan Johnson used the previous allegory to explain how ideologies and the words we use to express what we believe are what actually separate us, not our destiny or our quest. He peppered his conversations with anecdotes of meetings he held with Sufi, Hindu, Buddhist, Christian, Jewish, and Taoist world leaders to point out how these mystics—people whose lives are dedicated to the search for transcendent truths and the practice of spiritual disciplines—communicated. They basically spoke the same language, while many of their own followers, who were perhaps still far from the light, disputed—sometimes to death—the validity and exclusiveness of their dogmas. Separation is not just an illusion; it's a false belief.

What would the world be like if instead of arguing who is right, we tried to learn from each other or explore what we have in common in order to build knowledge together and further advance as humanity?

Let me share with you one of the experiences I had that contributed to my awareness that we are all connected.

One day, about fifteen years ago, I entered into a kind of trance or meditative state while attending an event with Father Richard McAlear, a charismatic priest who offers healing masses. His ministry postulates that for us to heal, wholeness needs to be restored, and he offers mental and emotional healing through prayer. Motivated by the stories I'd heard about his events, I went to see for myself. I didn't know much about the charismatic movement, but everything involving the laying on of hands or energetic healing seems intriguing to me as a Reiki master. After receiving communion, in the second part of the event, people lined up and approached the priest so that they would be anointed. Many of them fell backwards as soon as McAlear put his finger on their foreheads (there was someone behind each person to prevent them from hitting the floor). Suddenly, as I watched from my bench, I felt enveloped in a dust or golden light that emanated from all of us or was falling on everyone, and I was overwhelmed by an emotion that I can only describe as love. I'd lived many years with a consciousness of a clear but arbitrary line existing between *them* and *us*, and I had echoed that illusion of separation that leads us to repeatedly compare ourselves with our neighbor, and which makes us look at the other with distrust and judgement. But at that moment, I had a sudden sensation that I can better describe by paraphrasing the British philosopher Alan Watts:[19] The others and I were all as much continuous with the physical universe as a wave is continuous with the ocean.

When superfluous differences were erased, even momentarily, I was certain that no matter where others came from, no matter their skin color, ethnicity, origin, or gender, we were one humanity sharing the same tribulations and dreams. Unfortunately, it's not easy to maintain such awareness at all times.

When you breathe, trillions of air molecules fill your lungs, even though you will only use a small fraction of the oxygen contained in that air. However, that oxygen is essential for you to live, and it connects you to the whole planet. Where does this oxygen come from? Remember that most living organisms use oxygen. It's also safe to say that after we humans share a space, part of the oxygen and the carbon dioxide in the air I breathe comes from your body and will become part of mine when it's utilized.

As a Reiki practitioner and student of other vibrational therapies, I've had many opportunities to perceive the existence of energy (or electromagnetic) fields in the human body. Reiki practitioners learn what in Japanese is called Byosen Reikan Ho—usually translated as a technique that uses the hands to scan the "energy impressions" of an illness.

At the end of the day, the truth is we are all made of the same type of atoms, whose electrons exhibit properties of both waves and particles.

---

[19] Watts was a major contributor in spreading Eastern philosophies in the West. Watch: "You're It," at youtube.com

The body is electric and generates a magnetic field. Can we perceive the electromagnetic energy coming from the atoms' activity? Likely.

But beyond this experience of interconnectedness (one of several) that helped me change perspectives and made me aware that separation was an illusion, I would not have engaged in this book project if I had not also found so many loving people on my path as well as so much scientific evidence that we are born hardwired for bonding and responding empathically to others. I strongly believe that given a favorable environment we can keep an awareness of connectedness and consciously develop our capacity for empathy, compassion, love, and solidarity. Think that all we can offer the world is a future. I am, however, well aware of the obstacles we face in learning and practicing love.

With all of our potential as human beings, humanity as a whole is not doing things right. We are committing millions of costly and probably irredeemable acts that cause heartbreak or elicit hatred on a daily basis. This explains why I write with such urgency, seeing that humanity is going through a serious crisis that's causing immense suffering to millions of people and is not only demoralizing us but could even destroy us as a species because it's also severely damaging the planet. I believe all of the tribulations overwhelming *Homo sapiens* could be relieved by developing empathy and compassion.

We have much to learn from other cultures and ideologies. Taoists, for example, believe that we're born as loving beings. Using a dialectical vision of the world, they explain development of natural phenomena as the product of the movement generated by the interaction of opposing forces in nature (the theory of Tao—all—and yin and yang polarities). When this thinking is applied to medicine, it offers a vision of a multidimensional body that some might call idealistic, magical, or symbolic, but that constitutes a dynamic and beautiful explanatory model of a compassionate, peaceful, joyful, and wise nature of human beings.

According to Qi Gong—ancient practice of traditional Chinese medicine—at the moment of conception five lights (agents), Yellow, Blue, Red, White, and Green, escort the human soul (Yuan Shen[20]) toward the newly formed embryo. Each agent comes to inhabit different organs and imbues the person with certain primordial virtues (or congenital emotions). Each light is associated with one or more spirits and its corresponding sounds, colors, and elements (earth, water, fire,

[20] To the Chinese, this concept of soul has a different connotation than the usual given by major religions.

metal, and wood). In the course of life, and because of our experiences—especially traumatic ones—the lights become turbid, and new emotions displace the original ones. According to medical Qi Gong practitioners, the new emotions coming from life experiences end up clouding the original lights, giving origin to physical and mental symptoms. Their treatments consist, therefore, in purging toxicity (eliminating turbid light) to restore both the free flow of energy and the original lights and emotions. In this belief system, the green light is associated with the wood element, representing the growth of matter. The corresponding organ is the liver. The spirit inhabiting the liver is Hun, which makes the psyche move toward others in human relationships. The congenital emotions associated with the liver are love and compassion. Abandonment, rejection, or betrayal can lead to acquiring emotions such as anger, hatred, and resentment, which, if nesting in the liver and gallbladder, cause symptoms and dis-ease. Some traumas could displace the Hun spirit, with the consequence that the person would be separated from their conscience. The medical Qi Gong practitioner can intervene to retrieve the soul (with its memory of congenital emotions of love, compassion, unity) and prescribe treatments that include visualizations, uttering sounds, breathing practices, and specific movements to complement the process of purging and toning. This explanatory model, which sees the whole, the interconnections, the multidimensionality of the body, seems natural and integrative to me.

## Civilized?

When we speak of civilization, we refer to our intellectual, scientific, and technological achievements, but in the process of becoming civilized, we've built a society in which we've lost ourselves in the haze of appearance, in the illusion of owning, be it knowledge, status, or stuff. We haven't learned or have forgotten how to tend bridges between us. It amazes me that we have the audacity to call this society civilized and use this term as a synonym for *superior*, while we haven't found the ways to break the barriers that separate us. A significant part of our lives is very programmed. We relate with impersonal correctness, and we're short on critical thinking. We seldom fully embrace our common humanity or commit to our common destiny with full responsibility.

The social and political polarization taking place in the United States (and elsewhere) doesn't reflect the way the country traditionally defines itself and, of course, doesn't translate into love of neighbor.

Branded as progress, modernity splits us and freezes our relationships. The destruction of the planet happens while we live doped by TV, cell phones, the internet, drugs, or alcohol and while we lead an existence that's more automatic than fully awake.

A society that once held high standards of professional and moral behavior has become a culture where envy, jealousy, and indifference toward, or contempt for, others reign. Critical thinking and objecting to unacceptable behavior or political corruption is one thing, but it's something else to demean others just because they look different or don't think or act like us.

We will rightfully call ourselves civilized the moment we start being authentic and expressing and nourishing our capacity for love. No more competing and expecting to be rewarded just for doing the right thing. No more being dazzled by the glitz offered by a consumerist society. No more being unwary followers of false prophets promising paradise while unethically pursuing their selfish agendas.

Building meaning in life requires persistent work. The reward is that once love and compassion become our guiding principles, we'll be able to build and strengthen supportive societies committed to the preservation of the planet. We'll also be able to find nonviolent ways to resolve conflict, prevent injustice, protect the earth and, in short, rebuild a better world.

The struggle among classes, the hierarchies, the restriction of rights of the many in benefit of the privileges of the few, has characterized the world at least since we have recorded history (roughly five thousand years).

In the strict sense of the word, *civilization* (Egyptian, Mayan, Aztec, Inca) is used to define a society of complex organizations and advanced technologies, institutions, and social structures that has overcome great obstacles, that reveres intelligence, and rules over nature.

However, in the twenty-first century, when even scientific thought seems under siege, the most lauded brains are perhaps those of the wise guys who, in a stroke of luck, win millions of dollars gambling in the stock market or deceitfully declaring bankruptcy to avoid paying creditors. These days, many people often avoid having serious discussions about essential issues because very often discussions are framed by appeals to emotion instead of being a healthy intellectual exercise of individuals searching for the truth. It often seems that the social imperative is to shut up and fit in and just be very careful to not offend others. This is a fearful, artificial version of what tolerance and respect for diversity should be. We opt out of debate while at the same time we penalize dissenters with isolation.

In a speech before the Lowry Institute, Bret Stephens, columnist for the *New York Times*, defended the art of disagreement in this way:

> To say the words, "I agree"—whether it's agreeing to join an organization, or submit to a political authority, or subscribe to a religious faith—may be the basis of every community.
>
> But to say, I disagree; I refuse; you're wrong; *etiam si omnes—ego non*—these are the words that define our individuality, give us our freedom, enjoin our tolerance, enlarge our perspectives, seize our attention, energize our progress, make our democracies real, and give hope and courage to oppressed people everywhere. Galileo and Darwin; Mandela, Havel, and Liu Xiaobo; Rosa Parks, and Natan Sharansky—such are the ranks of those who disagree.[21]

A related problem is the rise of a digital culture, in which we're reading less and less literature and history, said the late academic James R. Flynn. A specialist in political studies, Flynn demonstrated in the nineties that environment rather than genes determines the IQ. As soon as modernization made education more accessible to a larger population, the average IQs progressively increased (except in the last twenty years, according to researchers from the Ragnar Frisch Centre for Economic Research in Norway). This is known as the Flynn effect. He suggested that an increase in the intelligent quotient was actually an increase in abstract problem solving rather than intelligence.

The new generation, that of the so-called millennials, tends to live in the present, with little historical perspective, Flynn has said. Therefore, it must be difficult for them to anticipate the future. This is dangerous. It's difficult to live up to what the historical moment demands or be better citizens if we are unaware of where we come from.

So, we're wrongly using the term *civilized* to boast of a certain material and intellectual development and a dominion over nature, while we've lost sight of the consequences of our accomplishments. And in that sense, we'll be known tomorrow as the civilization of urbanization and bulldozers (which devastate nature), road levelers, excavators, and prefabricated houses. We are the civilization of plastic—which is wreaking havoc on our oceans—and of electric cars, which could perhaps replace vehicles that today run on fossil fuels. We will be cataloged as the era of transnational corporations, of consumerist, monopolistic, and post-imperial capitalism; of mass education, free trade agreements, neoliberalism, and globalization of the economy. We'll also be remembered as a world that gave rise to terrorism, a sedentary

---

[21] B. Stephen, "The Dying Art of Disagreement," *New York Times* (September 2017).

lifestyle, refined sugar, cancer, coronary heart disease, gluten intolerance, and metabolic syndrome.

This is a time when digital communication and the extension of the scope that characterized social, political, and legal institutions to the international level (globalization) allowed easier commercialization of surpluses (due to overproduction), a typical problem of a laisser-faire economy, which is still mired in the fallacy of a market regulated by supply and demand. More sadly, our time will be known as one in which the trade of all kinds of weapons proliferated and the access of civilian populations to automatic rifles made awful crimes possible. Moreover, this is an era that will be remembered for senseless massacres perpetrated by fourteen- and fifteen-year-old children, a time in which local wars multiplied, and devastation by war widened, while a culture of consumerism, necessary to guarantee the marketing and consumption of goods, spread, thanks to the psychology of advertising.

Civilization? When egocentricity and a lack of empathy, compassion, and solidarity are rampant?

Few tribal societies remain on the planet from which we could relearn how to stay in close connection with nature. Those exploring the ancestral wisdom of these few tribes often misinterpret or trivialize the content of the shamanic traditions and the essence of their rituals. This sometimes leads to what might be called "spirituality for sale." For example, ayahuasca (also called yagé), a beverage made from herbs traditionally used in ceremonies and magical practices among the indigenous groups of the Amazon basin, has become fashionable in the United States as a panacea-drug (mixed with *Psychotria viridis*, a hallucinogenic plant containing dimethyltryptamine). Presented as a healing marvel, said to expand awareness, ayahuasca has gained social and legal acceptability.

The institutions supporting the status quo rarely, if ever, focus on promoting harmony or the nonviolent resolution of conflicts. Proclamations of a "government of the people by the people and for the people" have a hollow ring when the government engages in new forms of oppression, surveillance, and control.

I write at a time when information and knowledge generated by thinkers and scientists around the world are literally within reach of any person with access to a mobile phone or a computer. However, it's also a time when great powers conspire to privatize the internet and confiscate freedom of expression and association, subtly but systematically. It is indeed a time of paranoia and the proliferation of conspiracy theories fueled by a growing threat to privacy. We live in times of mandatory vaccines, many of them unnecessary and problematic, of instruments for facial recognition, street cameras, chips under the skin. The latest technological advances make it possible for an autocrat to take absolute control. The Snapchat app (or FamiSafe or Glympse, or others) allows users to track another person's whereabouts. A bit of imagination is enough to anticipate the uses of this technology in conjunction with the cameras installed in public places and ATMs, the cloud that stores our private information, or affective computing (devices will recognize, interpret, process, and simulate our emotions to tailor content offered to us).

This doesn't seem a favorable time for self-esteem or love of neighbor. Instead, this is an era of polarization and sectarianism when globalization exposes us to news about possible threats to our safety, and our fight-or-flight circuits, associated with aggressive and territorial behavior, are constantly aroused. And we respond by rejecting other cultures and customs. In an era when survival is more competitive, our reward system is also easily stimulated, which might explain a tendency to take refuge in the virtual world of social media, video games, or pornography.

In the midst of turmoil, and to gain a sense of security, we cling to beliefs, values, and customs induced by a selective grouping mechanism that builds protective barriers and an illusion of identity. We're coming together not to create attachments but to protect ourselves. As connecting strategies, we use plots, polarization, and animosity.

This is not a civilized time but a time of tremendous fear, distortion of information, disrespect, and intolerance. We seem trapped on a road without exit and paved with hate, a road generating violence, which leads to more hatred, and an even broader extension of violence.

## Laughter: bullying bullies?

In the atmosphere of separation, confrontation, sectarianism, and violence described in these pages, bullying has become a widespread occurrence. Statistics show that in the United States, at least 70 percent of children have witnessed bullying in schools, and 30 percent have been its victims. These numbers are similar to those reported in Colombia by the Departamento Administrativo Nacional de Estadística—DANE (National Administrative Department of Statistics). In Spain, the percentage of reported bullied children is 10 percent, with variations by region. In a comparative study of sixty-six countries, sponsored by Denmark, Tajikistan is the only nation with a prevalence of bullying of less than 20 percent (7.1 percent).

But bullying is not limited to schools. It arises at work, within the family, and more recently on the internet.

In October of 2012, Amanda Michelle Todd, a Canadian teenager, committed suicide after producing a video in which she shared her desolation as a victim of cyberbullying. Her case sounded an alarm for the whole world, which then began to pay closer attention. The consequences of bullying can be devastating for a young person. Bullying victims are two to nine times more likely to report suicidal thoughts, according to a 2008 Yale School of Medicine analysis.

Where are the regulations, reforms, laws, or public debates on the subject? Where is the public consciousness that rises collectively against this form of abuse? Although we're more aware of the existence of bullying, and even though a few new laws[22] have been passed, the culture that allows and sometimes encourages harassment has not yet changed.

It doesn't set a positive precedent that the United States has elected a president who has fostered prejudice, discrimination, and violent behaviors, with some media echoing his conduct. And it's not favorable either that his bravado has become standard behavior when dealing with crucial national and international issues. We have come to a point where many people customarily avoid the news and the media because it leaves them exhausted. I'm among those who opt to follow what's going on by listening to comedians who pick up on the news. At least they have created a circus out of this character, and when presented with humor, current national and international matters seem less stressful.

The founder of psychoanalysis, Sigmund Freud, considered jokes the product of the activity of the subconscious mind.[23] Certain feelings of hostility, aggression, or sexuality can be expressed with humor, bypassing the mind's censors. The joke usually turns its subject into a target onto whom we safely release contained hostility. However, aren't jokes and comedy very often another bullying device?

In Orrin Klapp's book *Heroes, Villains, and Fools* the author states that societies find fools who provide people with relief from routine and discipline. Comedians can touch on taboo material. But doesn't that also desensitize us and make us less able to sharpen our mind and critique moral and social issues that affect society? Does it make us less likely to take action?

Even though comedians sometimes denounce wrongs and help us release tensions, do jokes contribute to generating change? The comedians

---

[22] See: www.stopbullying.gov/laws/index.html

[23] Jokes are a playful judgment, said Kuno Fisher, philosopher, historian, and critic on whom Freud based his work on the topic of jokes.

make us laugh at the expense of a person or tyrant portrayed as evil, ridiculous, or inappropriate. And we laugh together as a group and feel relieved, and we not only accept this other type of bullying, but the paradox is we condone it even when it's used to condemn some bully's behavior.

I don't deny that some comedians play a social role denouncing and critiquing different forms of abuse of power, but they risk trivializing serious matters. The audience identifies with the comedian, and laughter is cathartic. The downside is that criticism, judgment, the reproof behind the showman's monologues seldom turn into action in pursuit of solutions. Even though they could help to creating awareness, comedians risk contributing to perpetuating the status quo. They feed our cynicism and despair when what the world most needs is optimism (or *possibilism*, as researcher and author Frances Moore Lappé would say). We need some certainty that we will be able to find solutions, that these are possible and that we can all contribute to putting them into action.

Isolation, ridicule, and other forms of emotional, verbal, and physical harassment that characterize bullying, cause suffering. Nothing is farther from solidarity, compassion, and love than a culture fostering animosity and contempt. The existence of bullying, from grade school to the presidency of the United States, is a bad symptom of the ills that plague humanity. Bullying extending from the individual level to international relations is validated as an apparently innocuous charade.

The evils affecting us are not going to be cured by laughter. A constructive element of humor is its potential to generate embarrassment and thus, change behaviors, but this is only possible when the target has a capacity for self-awareness. On the other hand, it's worth pondering to what extent previously reprehensible behaviors become acceptable and even funny (the behavior of a drunk in public, for example, or even the humorous interpretation of a president of the United States in shows like *Saturday Night Live*), fostering or even inciting the standardization of certain behaviors instead of having a critical function.

## PART TWO: Obstacles to Love

*World peace can be achieved when the power of love replaces the love of power.*

—Sri Chinmoy

There are various obstacles to learning love and determining whether or not we can approach others with compassion. Some obstacles are of an interpersonal nature (having to do with the environment, the family, or the community in which we grew up), and others are intrapersonal (dependent on the way we process information, perceive the world or adopt a set of beliefs). One of the biggest obstacles is the fact that we live in a world of stratified social organizations and economic systems characterized by an excessive ambition for power. More often than not, our leaders perceive war (or the threat of war) as the best option for resolving conflict. Repression trumps persuasion in stratified societies, and profit becomes more important than the well-being of the people or the preservation of the planet. Social injustice is the outcome. This is not an opinion; it's a fact.

In a world organized around power relationships between parents and children, brothers and sisters, friends and peers, teachers and students, humility and vulnerability might be interpreted as deplorable weaknesses.

How could a culture of solidarity flourish when the most vulnerable, instead of being supported, are rejected and chastened? And what is the origin of this type of interaction? I believe social inequality is at the root of the problem.

The terms "horizontal plane of movement" and "vertical plane of movement" were introduced by the brilliant Austrian doctor and professor Lydia Sicher in 1955 to illustrate ideas presented by the psychoanalyst Alfred Adler, who asserted that the essence of mental health was the ability to connect with others. In Adlerian terms, the term horizontal plane of movement refers to a social organization with a sense of equality in which all share the goal of solving problems in cooperative ways: "people striving for superiority in line with social interest" (Sicher). The term vertical movement plane refers to an effort to achieve personal superiority, which ultimately leads to different forms of domination. Because it is not in line with the harmony of the group, it leads to deep levels of discouragement.

The horizontal plane of movement implies an equitable way of operating, and it must lead to democracy. A person operating in this plane asks, "What does this situation require?" Conversely, operating in a vertical plane is an egocentric way of social movement in which individuals climb a ladder in search for status, which leads to a hierarchical type or organization. The question of the one who operates on this plane is, "How do I look?" Or, "How am I doing?"

I was raised by a Catholic family and educated in a Catholic school for girls. I grew up in an environment in which religion played a value-forming role, so it's not surprising that Jesus's message, "Love thy neighbor," was presented to me at a very early age, not as a suggestion but a commandment, which I took very seriously without being aware of the magnitude of the commitment I was making.

By the end of my medical school years, I had met René García, a Colombian priest who preached the "theology of liberation." Oppression is not an act of love, he said, and therefore, "it is against Jesus's commandment." Becoming a good Christian demanded siding with the poor.

García told us about the 1968 Latin American Bishops' conference in Medellín. The Bishops declared that poverty was the result of industrialized nations profiting at the expense of developing countries.

His ideas sowed an uneasiness with inequality in me. García's words became an invitation to look at a more dialectical understanding of the world. I became disapproving of abusive relationships between countries, within countries, or between institutions or people. I bristled at the idea that we owed the World Bank and the International Monetary Fund an unbelievable amount of money and that loans in the seventies were the most onerous as yet imposed on poor countries.

In 1971, my father stopped growing wheat. That year the United States had found a way to persuade some Latin American governments, including Colombia's, to buy their surpluses, which had been rejected by China (wheat production in the Far East was soaring) and Russia. Colombia's mills received subsidies from the government for buying US wheat, while the nationals were left with no buyers for their product. This caused the price of wheat produced in the country's farmlands to fall alarmingly, leading to the ruin of thousands of producers. How could that be?

In the 1970s, I witnessed protests against the Population Council of the United States, which was performing systematic, unconsented sterilizations (tubal ligations) of women living in rural populations in

Colombia, Bolivia, Puerto Rico, and Peru. The Council was also offering incentives (transistor radios was one I remember) to men who agreed to have vasectomies. Even more enticements (like free courses and surgical instruments) were offered to rural doctors willing to receiving training in how to do the tubal ligations. Instead of using persuasion and educating people about contraceptive methods, the Population Council forced its policies on a vulnerable population with the argument that they knew better what suited us.

In the eighties, the foreign debt exceeded the earning power of most countries, which led to what is known as the Latin America debt crisis.[24] Countries followed Mexico's resolve not to cancel the debt. My eyes were wide open then. I wanted to know more, do more.

I also witnessed multinational companies in collusion with the Colombian government, progressively appropriating the oil fields and coal mines (such as El Cerrejón mine) by entering into disadvantageous contracts. In places like Monte Líbano, Colombia, instead of our national army, our government allowed that US Marines patrolled the camps where the BHP Billiton and Hanna Mining Company personnel lived while working for the Cerromatoso ferronickel mines and plants. For decades, mining in Monte Líbano caused heavy-metal pollution health problems affecting mostly indigenous people (but pollution was only verified in 2015, four decades after mining had started).

In Colombia, exploitation of natural resources has intensified lately, thanks to the development of new technologies and the arrival of new foreign investors. The mining has devastated forests, dried rivers, and polluted water sources, threatening the survival of the adjacent geographical areas and populations.

In the past three decades, about 70 percent of the Peruvian Amazon rainforest has been sold to multinational companies for oil, gas, and mining operations. The indigenous people who live in these jungles were not consulted. The Amazon rainforest, the lungs of the world, have been burning. The Brazilian Instituto Nacional de Pesquisas Espaciais (National Institute for Space Research) said a record number (nearly 73,000 fires) had been detected in 2019. Many blame the deforestation, the goldmining, and the reversal of environmental policies by current Brazilian president, Jair Bolsonaro.

Along with the growth of the North American industrial presence in Latin America came its advertising, television, and films, presenting the "American Dream" as an idealized model of the life to be pursued.

---

[24] See the Federal Reserve History article "Latin America Debt Crisis of the 1980s," online at Federalreservehistory.org.

Cultural penetration also came through comic books, which became agents of socialization, modeling social values and gender roles. I think it was the Mexican cartoonist Rius who made me realize that the characters appearing in TV cartoons and newspaper comic strips (many of them translated into Spanish and very popular in Latin American countries) displayed certain types of either patriarchal or atypical family models. Men were always the breadwinners and women the homemakers, for example. A nuclear family was often either nonexistent, dysfunctional, or quite conflictive. Do you remember Mandrake and Tarzan with their lifelong girlfriends, or were they lovers? Both Donald Duck and Mickey Mouse had nephews, and their girlfriends had nieces. The relationships never seemed to progress to the formation of a whole family. And who were the parents of those nephews and nieces? Female characters were rather angry women (who had to educate their men) or vain and shallow girls. Male figures were kind of grumpy or good-for-nothing (think Dagwood Bumstead). The relationship between Tom and Jerry, between Elmer and Bugs Bunny, between Dagwood and his boss, perpetuated stereotypes of males as either smart alecks, angry, or lazy types.

I don't think we have studied how these unloving models of family and relationships were implanted in our minds and lifestyles through comics, television, and films or how they affected who we eventually became. However, I'm aware of the stereotypes I assimilated through them.

My frustration with the state of affairs in the world grew fast in a very turbulent and violent world. The news told us about the Soviet invasion of Czechoslovakia in 1968, the same year in which US intervention in Vietnam peaked and Israel invaded the Gaza Strip. The USSR invaded Afghanistan ten years later. Israel and Palestine got engaged in what has become an endless war. When the USSR asked Cuba to send troops into Angola, Fidel Castro sent ten thousand reluctant soldiers. With the threat of nuclear war between the two superpowers (the URSS and the USA) always in the air, we wondered to which refuge we would run to protect ourselves. Over the course of the next few decades, several Latin American dictators rose or fell with the acquiescence or military intervention of the United States government.

In response to all this turbulence, the call to contribute to building a better world became a hallmark of my generation, from the hippies making the symbol of peace and saying no to war to the extremist revolutionaries (guerrillas) that plagued countries like Colombia.

My main focus became exploring ways to be of service to the needy and to build solidarity between workers and between good people. New insights made me reevaluate family relationships, love relationships, and of course, the purpose and meaning of our lives.

## The pursuit of happiness

*The purpose of life is not to be happy. It is to be useful, to be honorable, to be compassionate, to have it make some difference that you have lived and lived well.*

—Ralph Waldo Emerson

To develop a holistic, multidimensional approach to the complex factors that either favor or hinder the process of learning love would require us to discuss the current state of affairs in the world, but that would be topic for another book.

Let me admit that, in what I'm about to say, I'm reacting to those authors who, talking about our very intricate world issues and uncertain future, make the unabashedly optimistic assertion that we've never had more reasons to be happy than we do now. Technological advances leading to less suffering, a higher life expectancy, and lower infant mortality are without question positive, desirable, advances. However, access to comforts that were unimaginable 150 years ago, amenities such as jacuzzi bathtubs, a Bugatti car, or a Goldstriker iPhone Supreme might give us only ephemeral joys, not happiness. The United Nations defines happiness as a state of mental or emotional well-being characterized by positive or pleasant emotions that range from joy to intense joy.

Although economists have found a correlation between income and happiness, two Nobel prize winners (economics), Angus Deaton (2015) and Daniel Kahneman (2002), have published research in which they agree that such correlation stands true only for people with an income up to $75,000 per year. (Bear in mind that according to the Pew Research Center, only seven percent of the population makes above $50,000 a year.) Emotional well-being does not increase above that level of income, although life satisfaction may increase.

Princeton University Center for Health and Well-being scholars Daniel Kahneman and Angus Deaton analyzed the responses of 450,000 residents to the 2008–2009 Gallup-Healthways Well-being Index.[25] The survey considers two aspects of well-being: emotional well-being and

---

[25] D. Kahneman and A. Deaton, "High Income Improves Evaluation of Life but Not Emotional Well-Being," *PNAS* 107 (38) (Sept. 21, 2010): 16489-16493. doi.org/10.1073/pnas.101149210.

life evaluation. The first refers to the subjective emotional experience and the second to the thoughts people have about their lives.

"People's emotional well-being is constrained by other factors, such as temperament, sense of purpose, and life circumstances," the researchers said, and also concluded that, "The pain of life's misfortunes, including disease, divorce, and being alone, is exacerbated by poverty."

Even if life satisfaction increases steadily with income, it's noteworthy that among about 150 countries, the US was worse on worry (eighty-ninth from best), sadness (sixty-ninth from best), and anger (seventy-fifth from best).

In my observation, people who have developed a healthy relationship with themselves (self-love), with others (love of neighbor), and with their environment (love of the planet) experience joy more often than most of those who have millions of dollars or an abundance of material possessions. The ability to experience or cultivate joy comes from an increased consciousness and positive outlook of life. It comes from living in the present, having realistic aspirations, appreciating what one has (I didn't say owns), and finding satisfaction in altruistic actions. It comes from inner harmony, knowing who we truly are, accepting life as it is.

English sociologist and epidemiologist Richard Wilkinson, with Kate Pickett, authored *The Spirit Level: Why Greater Equality Makes Societies Stronger* (2009, Allen Lane), among writing other books on inequality and unhappiness. Wilkinson is a cofounder of the Equality Trust, and he says it's a myth that societies with high accumulations of wealth are happier, more loving, or more peaceful. Wilkinson argues that the greater the gap in the distribution of wealth is, the less happy a nation's people are (including those who have wealth), and the less healthy and trusting. Indicators used by Wilkinson included infant mortality, drug abuse, mental health problems, obesity, homicides, and school dropout rates. These indicators worsen (as does social inequality) in countries with the greatest gaps between rich and poor,[26] like the United States, Singapore, and England, while countries like Japan and Norway, with smaller wealth gaps, fall at the other end of the spectrum.

Other studies show that social inequality causes great stress, which affects physical and mental health. In very stratified societies, hierarchies increase competition and raises the bar on what you're supposed to achieve, which in turn increases the levels of distress.

Jigme Singye Wangchuck, King of Bhutan (a small Buddhist country) proposed in 1972 a new indicator he called Gross National

---

[26] The gap is measured by comparing the richest 20% with the poorest 20% of the population.

Happiness (GNH) to measure quality of life. Tired of criticism from other nations about the economic situation in his country, the king instructed a planning commission to measure the level of joy and comfort among the inhabitants. He wanted to make sure that prosperity ran parallel with a good government, preserving traditions, and protecting the environment. The King pointed out that conventional models used the gross domestic product (GDP) as main indicator, while the true measure of the development of a society did not reside in its economic performance but in a combination of both material development and spiritual advancement. The Bhutanese GNH sits on four pillars: good governance, preservation and promotion of culture, environmental conservation, and sustainable and equitable socio-economic development. The four pillars were further elaborated into nine domains: living standards, education, health, environment, community vitality, time-use, psychological well-being, good governance, and cultural resilience and promotion.

In 2007, for the first time, the European Commission considered replacing or complementing the GDP as a sole measure of progress with a happiness index, as had already been done by other countries, including Brazil, Thailand, and Canada. There is now a tendency for nations to consider its people's happiness as a true measure of progress and establishing it as a public policy goal. The first World Happiness Report was published by the Earth Institute following the 2012 United Nations high level meeting on happiness and well-being. A happiness index was established, based on six indicators: freedom (from oppression), social support, life expectancy, income, generosity, and a trustworthy government. The UN Assembly proclaimed March 20th as the International Day of Happiness.

In the 2019 reports, the Nordic countries were at the top of the list of happy countries (Norway leads, followed by Denmark, Iceland, and Switzerland), while the United States, ranks at number 19, and happiness rates dropped five positions since 2017. Despite improved income, life expectancy in the US has dropped, while the social indicators (social support, generosity, perceived corruption of government and business, and sense of personal freedom) deteriorated. The Organization for Economic Cooperation and Development (OECD) also reports that the United States subjective wellbeing has dropped. In 2015, the GDP in China had multiplied more than five-fold over the previous quarter century, but subjective well-being over the same period fell for fifteen years before starting a recovery process.[27]

[27] See the World Happiness report at: worldhappiness.report/ed/2017/

It's not surprising that the location of one of the bloodiest modern wars, Syria, is 152 on the list of the 155 countries included in the happiness report. African countries lag behind all others.

## The fierce god of war

*A prince should therefore have no other aim or thought, nor take up any other thing for his study but war and its organization and discipline, for that is the only art that is necessary to one who commands.*
—Niccolo dei Machiavelli

In his book *The Better Angels of our Nature: Why Violence has Declined*, Harvard psychology professor Steven Pinker states that the incidence of violent acts has significantly dropped worldwide in the last two hundred years, even if we include all wars, armed conflicts, and terrorism. Pinker argues that humans become less violent when they become more civilized—when they accumulate greater wealth (to make life more pleasant) and create laws (forming democracies that are inherently repressive, so-called law-and-order states). Pinker believes that in the past few decades, although there have been more civil conflicts, we see fewer wars between large powers or between states (which cause more deaths). Barbaric practices, he says, such as human sacrifice, executions, and torture have been abolished. Actually, we know that execution and torture are still used. In 2014, Amnesty International found evidence of torture practices in seventy-nine countries, all of which were part of the group of 155 countries that had ratified the United Nations convention against torture. But Pinker may be right when he says there is less cruelty against women, children, and animals.

Pinker believes the invention of the printing press (which facilitated education), more recently of the internet (which facilitated the dissemination of information), the rise women's liberation movements, and an expansion of compassion as a social value have contributed to the creation of a more empathetic world.

However, his views on the decline of wars are controversial. Other scholars, such as professor, author, and specialist in international relations, Dr. John Arquilla,[28] disagree with Pinker. Even though since

[28] J. Arquilla, "The Big Kill—Sorry, Steven Pinker, the World Isn't Getting Less Violent," *Foreign Policy* (December 3, 2015): https://bit.ly/24czMLU.

the very bloody Second World War, the great powers have not had direct confrontations, they have participated in many wars indirectly. Examples include the Soviet invasions of Hungary, Czechoslovakia, and Afghanistan; the Vietnam War; the Iran-Iraq War; the military support provided by the United States, Russia, and the United Nations to either the government or the insurgents in Syria; and more. And although war has changed in form, it's not less destructive than before. Just look at the images of the terrible devastation in Syria, for example.

In *The Soul of the Marionette: A Short Inquiry into Human Freedom*, (Allen Lane, 2016) John Gray points out that Pinker's stats are misleading:

> While it is true that war has changed, it has not become less destructive. Rather than a contest between well-organized states that can at some point negotiate peace, it is now more often a many-sided conflict in fractured or collapsed states that no one has the power to end. The protagonists are armed irregulars, some of them killing and being killed for the sake of an idea or faith, others from fear or a desire for revenge, and yet others from the world's swelling armies of mercenaries, who fight for profit. For all of them, attacks on civilian populations have become normal. The ferocious conflict in Syria, in which methodical starvation and the systematic destruction of urban environments are deployed as strategies, is an example of this type of warfare.[29]

Another thought is that the civilian population has been more affected in the last few decades than in any previous times, except for the Holocaust during the Second World War. The conflict in Syria has caused more than half a million deaths, and children die of hunger in besieged cities, while urban centers are systematically destroyed. The fact that the terrorists target defenseless civilian populations is another terrible example of the ferocity of current conflicts.

The consequences of these wars include exodus, refugee camps, and famines.

In 2015, twenty-four people were displaced every minute. By the end of 2018, there were close to seventy-one million displaced people (the highest figure ever registered and half of them children), twenty-six million refugees, and ten million stateless people. By the end of 2016, Europe had given asylum to at least seventeen million people, the United States to sixteen million, Asia and the Pacific to eleven million, the Middle East and North Africa to twenty-six million, and Africa to thirty

---

[29] John Gray, "Steven Pinker Is Wrong about Violence and War," *Guardian* (March 13, 2015): https://bit.ly/2mPaDrp.

million.[30] In 2019, Syrians were the top asylum seekers in the European Union.

Looking at the human costs of the conflicts that shake the world, it's difficult to see signs of what we could call civilization.

Machiavelli's influence is still felt in modern political thought. Probably his main contribution is a way of thinking that assumes humans are by nature selfish and self-aggrandizing, and a "prince" should focus only on the ends of his endeavors, irrespective of the means used. Putting aside moral considerations (characteristic of classical thinking) aimed at the creation of "good government," a ruler should primarily focus on examining more pragmatically what would constitute an effective government. In *The Prince* (1532), the diplomat provides advice on how to face enemies, using force and fraud in war. Machiavelli advocated the use of "power politics," and his contribution has since shaped international relations. He said, "Cities ordered for virtue will find their ruin at the hands of states organized for war."[31] His state model, which monarchies swiftly adopted, is behind many countries' imperial expansion in the past centuries.

If you pay close attention, you will see how often we all employ the narrative of war, which reflects the existence of a pervasively war-like culture. The way we conceive the world reflects a militaristic point of view—militarism as an ideology that advocates for force as source of safety.

Richard Nixon, for example, used the motto *Law and Order* to propel himself into the Oval Office, and so did Donald Trump. The true definition of a law-and-order state is one in which nobody is above the law rather than one that fosters an us-versus-them, hierarchical, racist, or nationalist mentality. Trump campaigned on the promise of the strong use of force, especially against terrorists, but also even against undocumented immigrants, some of which he described as rapists. Militarism assumes that power and safety are inextricably linked to the military. Humans have built societies that constantly consider and get ready for war. Just look at the budget allocated to the Department of Defense or the share of the budget that goes to any military-related expenses in the United States: military spending is around half of all federal discretionary spending.[32] Bear in mind that money allocated in

---

[30] Statistics from the United Nations High Commissioner for Refugees (UNHCR).

[31] M. Fischer, "Machiavelli's Theory of Foreign Politics" *Journal of Security Studies*, (December 24, 2007): 248–279.

[32] www.nationalpriorities.org.

the US budget for the military not only pays officers and soldiers and funds services to veterans, it also goes directly to paying war expenses and funding research focused on inventing more sophisticated and increasingly lethal weapons.

The United States spends more on national defense than China, Russia, Saudi Arabia, India, France, United Kingdom, and Japan combined. But the percentage of the total budget going to the military is not very different from that of the military budgets of nations such as Russia, Iraq, and Israel.

The 2018 Stockholm International Peace Research Institute (SIPRI) report on worldwide defense reveals that the United States leads the way in military expenditures, with a budget of $610 billion, accounting for 35 percent of military spending worldwide. China followed with an estimated $228 billion in defense spending. They were followed by Saudi Arabia ($69.4 billion), Russia ($66.3 billion), and India ($63.4 billion). These five nations collectively accounted for around 60 percent of the $1.74 trillion spent on the military around the world in 2017. Together, the top fifteen nations in defense spending in 2017—same countries as the year before—accounted for 80 percent of global defense spending.[33]

Another ironic example: Just after reaching the 2017's peace agreement with the guerrillas, Colombia increased its military spending, to the detriment of social programs, including education.

This militaristic mindset is expressed in more than just the money spent in weaponry, war, and defense. It permeates our everyday language. Military and sports metaphors prompt us to see everything in terms of conflict: we admire the "brave" and incite people to "fight" and "defeat" their circumstances. We praise with adjectives like "courage." We want to "triumph." We often use belligerent language and metaphors in our speech. This is particularly evident in the media coverage of political campaigns (as when they talk about "battleground states," or "ambushing an opponent," using a "secret weapon," or "dropping a bombshell.")

Even people with a progressive agenda might use expressions such as "battling for the soul of the party," "slashing" gas emissions and becoming "eco-warriors." When we talk about the body, we deem the immune system a "defense system," that responds to the "threats or attacks to the body," when a more accurate and current view is that of

---

[33] N. Tian, A. Fleurant, A. Kuimova, P. D. Wezeman, and S. T. Wezeman, "Trends in World Military Expenditure," *SIPRI* Fact Sheet (May 2018).

the physicist Fritjof Capra: instead of an army that chases, confronts, and destroys invaders, the immune system is a self-governing network that participates in the body's learning process and is responsible for its "molecular identity" and the biochemical communication between organs.

We need a new, multidimensional, paradigm to approach life.

This narrative of war—these linguistic choices we make—shape politics, education, sports, and the media. When parties feel their ranks are questioned, preemptive behaviors aimed at defending status can result.

Parenting and education in schools are often designed along the lines of dominant hierarchies, typical of the military, and this mentality could be costly for the physical, mental, emotional, and relational health of children.

Adults often say children have become "impossible" and require "stronger discipline." Or they demand their children "shut up" and "obey" authority. This viewpoint often prevents adults from seeing their own responsibility in what's happening with a child.

Children are rarely given the right to comment on decisions that affect them; they're told not to talk back, they must comply and behave according to rather inflexible norms (that follow the reigning idea that there's only one right way of doing things). Expressions like "we know what's best for you," "children are children," "discipline builds character," reflect this hierarchical us-versus-them mentality.

Parents and educators have tried for years to correct the behavior of children with physical punishment, timeouts, exclusion, deprivation of privileges (often recommended by their therapists). Not very different to how we treat criminal offenders, is it?

Still, there's strong evidence that both emotional and bodily pain inflicted by a caregiver interferes with the creation of secure attachments between parental figures and children and also affects the capacity of the child for emotional regulation.

If children "misbehave" very badly, they might be sent to a military school or a boot camp with the hope that a fairly rigid hierarchical structure and some hard work will "correct" their behavior.

Strong evidence suggests that high levels of stress generated by autocratic parenting or education styles affect children's brains, and therefore will not provide a fertile ground for the child's emotional development, much less for learning empathy. Repressive measures (confiscating their iPad or banning their favorite entertainment) would elicit nothing more than resentments and grudges that rise incrementally. Behavioral inhibition, —the child experiences discomfort or avoids being in the presence of people or environments that are not familiar—is often the end result.

The authoritarianism associated to a militaristic approach to parenting has also been correlated with rigid thinking (resistance to change and to seeking or enjoying new experiences) and can contribute to the development of obsessive behaviors. In addition, there is a negative correlation between psychological inflexibility and the development of intelligence and creativity, so essential to solving problems.

In *The Rise of Trump* (Amherst College Press, 2016), Matthew C. MacWilliams reports the results of a survey (of 1,800 people) done during the 2016 election. He found that four survey questions about childrearing were better predictors of who would vote for Trump than any other questions. Those who considered it very important to raise a child to be respectful, obedient, well-behaved, and well-mannered (expectations typical of an authoritarian parent) were more likely to vote for Trump than those who preferred to raise children to become independent, self-reliant, considerate, and curious.

In the past two or three decades, youngsters have had easy access to video games that simulate war (first video game—Spacewar—was released in 1962 by MIT).

"Military and other role-play items may help kids work through or cope with what is happening in the world around them through play rather than through outwardly aggressive behavior," the Toy Industry Association (TIA) publicly stated.

But studies show toy guns and video games do more harm than good, beyond just stimulating a militaristic mentality.

For example, in 2012, psychologists Brock Bastian[34] and colleagues scanned the brains of 106 undergraduates who had played either military games (like *Mortal Kombat*) or tennis and found evidence that playing the violent video game desensitized players to real-life violence and to the suffering of others.

Several other psychologists have asserted that typical first-person shooter, fighting and action games (like *Mortal Kombat* or *Call of Duty*) do desensitize children to the act of killing.

The reasons for manufacturing and marketing these games are not exactly innocent.

Video game manufacturers and firearms makers have a synergetic marketing relationship. In "Real and Virtual Firearms Nurture a

[34] B. Bastian, J. Jetten, and H. R. M. Radke, "Cyber-Dehumanization: Violent Video Game Play Diminishes Our Humanity," *Journal of Experimental Social Psychology* 48, Issue 2 (March 2012): 486–491.

Marketing Link" (December 2012), the *New York Times* reported that, according to several marketing experts, "Makers of firearms and related gear have come to see video games as a way to promote their brands to millions of potential customers."

Many of these war games are also sponsored by retailer companies such as GameStop that award large sums of money to promote gaming and offer expensive prizes to the best "virtual killers." In 2018, for example, *Dota 2,* a popular online battle arena video game developed by Valve Corporation, awarded a total of $41,326,628 to players participating in 117 different events. Let's remember that rewards lead to behavior reinforcement.

Studies that correlate the use of video games with violent behavior are still not conclusive (more studies are needed). But let's examine at least one lamentable case where such an association has been reported.

On April 20, 1999, two teenagers went on a shooting spree at Columbine High School, in Littleton, Colorado, killing twelve children and a teacher and wounding twenty-one other people before they committed suicide. Eric Harris and Dylan Klebold used semi-automatic rifles and homemade bombs. The murderous children were addicted to violent video games, and their attack was similar to those found on the game *Doom*. Harris had a created his version of the game, which he shared with Klebold, where the killer shoots people who are in no position to respond to the attack. Somehow, these boys transformed a wild fantasy into an appalling reality.

Lieutenant Colonel David Grossman was living in Arkansas on March 24, 1998 when thirteen-year-old Mitchell Johnson and eleven-year-old Andrew Golden perpetrated another massacre, at the Westside Middle School in Jonesboro. An expert in the psychology of killing and author of *On Killing: The Psychological Cost of Learning to Kill in War and Society* (Little, Brown and Company, 1995), Grossman said violent video games are "murder simulators," not very different from the simulators used to train soldiers in the art of killing the enemy. And video games could train children to kill. Grossman explains that killing requires training, because of our innate aversion to killing our own kind. Historian General Samuel Lyman Atwood devoted much of his life to analyzing conflicts from the First World War to Vietnam and documented the fact that as much as 75 percent of fighters (from different cultural backgrounds) refused to kill if they came face to face with the enemy.

The voices raised against violent video games are part of a complex controversy on what it would take to prevent more of these absurd killings. Only a small percentage of the killers have been identified as having mental health issues.

Experts consider it imperative not only to limit access to violent (war) video games but also to seek changes in the school culture, since bullying and retaliation promote and perpetuate violent behaviors.

In the twentieth century, mass school shootings killed fifty-five people and injured two-hundred-sixty others at schools in the US.[35] Just in the year 2019, by the time this book was completed, at least eight shootings had taken place in schools or colleges, resulting in sixty-six deaths and eighty-one injuries. Sixty percent of the killers were between eleven and eighteen years old. Several school shooters have carefully studied the attack perpetrated by Harris and Klebold or have cited them as role models.

Most of this century's mass shootings (Statista reports 62 percent) were perpetrated by white males. There are more public mass shootings in the United States than in any other country in the word.[36]

Whether or not video games contribute to the problem, violence is increasingly present in schools in the United States, and we need to ask why. In February 2018, Nikolas Cruz, an alumnus of Marjorie Stoneham Douglas High School, in Parkland, Florida, used a semi-automatic rifle to kill seventeen people in just a few minutes. Cruz was only nineteen, a young man with mental health issues who had been adopted as a baby, had lost his adoptive father when he was still a kid, and had lost his mother a few months before the killings. Cruz had participated in the Junior Reserve Officer Training Corps (JROTC), a youth organization of the US Army, financed in part by the National Rifle Association (NRA). JROTC teaches children thirteen and up how to use rifles. A combination of unresolved mental health problems, easy access to firearms, lack of regulation, a violent culture, and violent media influence make of this one a very complex case.

Attempts to regulate children's access to violent video games have not been successful in the United States. However, several countries in Europe and some North American companies have established video game age ratings.

In direct response to ratings by critics and viewers, TV and film producers are delivering an increasing number of shows and movies

---

[35] Springer. "Rapid Rise in Mass School Shootings in the United States Study Shows: Researchers Call for Action to Address Worrying Increase in the Number of Mass School Shootings in Past Two Decades," *ScienceDaily*. April 19, 2018): www.sciencedaily.com/releases /2018/04/180419131025.htm.

[36] A. Lankford, "Public Mass Shooters and Firearms: A Cross-National Study of 171 Countries," *Violence Victims* 31(2) (Epub January 28, 2016):187–99, doi: 10.1891/0886-6708.VV-D-15-00093.

depicting confrontation, war, and violence, and this is worrisome. In 2013, a report from the American Academy of Pediatrics (ACA)[37] stated that violence in movies has doubled since 1950, and violence in films rated PG-13 and up has tripled since 1985. Children and adolescents in the United States spend an average of seven and a half hours a day in front of a screen or listening to music. A good deal of the content of both forms of entertainment is violent. The Federal Communications Commission's 2007 report noted strong evidence of a correlation between the behavior of children and their exposure to violent images.

There is also evidence that images of violence have a detrimental impact on us, especially on a brain in formation. The changes observed in the brain are very similar to those registered when a person is facing real danger. Adolescents' brains are still going through synaptic pruning, a selective process that peaks when the child is between two and sixteen years old. During pruning, the axons, which transmit impulses between nerve cells, retract, particularly those that become unnecessary for cognitive processes, while the neuronal pathways most frequently used not only maintain their connections but become stronger. So, what is learned, what is repeated, what is experienced or witnessed, both at home and in schools, especially in the first years of life, becomes crucial for the learning process of the child and his or her ulterior behavior.

The American Association of Pediatricians (AAP) recommended that "pediatricians consider making children's 'media diets' an essential part of all well-child examinations," placing special emphasis on "guiding the content of media and not only limiting quantity."

A fMRI[38] study at the University of Indiana School of Medicine found significant alterations in the brains of young men who had been exposed to images of violence. The changes were more evident in the prefrontal cortex (thinking brain) and the amygdala (emotional brain) and were observed only a week after participating in violent video games.[39]

There is also a connection between firearm ownership and the occurrence of mass shootings. The Sunshine State, Florida, has been nicknamed by some the Gunshine State, not only because it's shaped

---

[37] B. J. Bushman, P. E. Jamieson, I. Weitz, and D. Romero, "Gun Violence Trends in Movies," *Pediatrics* Vol. 132. Issue 6 (Dec. 2013).

[38] fMRI stands for functional magnetic resonance imaging. MRI is not only a diagnostic tool, it has also facilitated the study of the brain and of behavior. Scientists study how our feelings or behaviors correlate with patterns of activity in different brain areas.

[39] "Violent Video Games Alter Brain Function in Young Men." *University of Indiana's* press release (April 3, 2018).

like a gun but because of the large number of firearms owned by the civilian population. In Florida, 32.5 percent of the population owns at least one gun. That's significantly higher than the national average of 29.1 percent. (Alaska has the highest rate of gun ownership, at 69 percent).

According to the same study cited above, "The United States and other nations with high firearm ownership rates may be particularly susceptible to future public mass shootings, even if they are relatively peaceful or mentally healthy."[40]

The lack of regulations limiting the access of young people and people with mental health disorders to guns is also a serious issue. Here's an example illustrating how heated the controversy on guns is: on March 2018, in an unprecedented move, Florida legislators passed a law restricting access to military grade weapons, including AR-15s, by people under the age of twenty-one. The powerful NRA responded swiftly, filing a lawsuit against the legislators contending that this measure went against the Second Amendment of the Constitution.

However, evidence abounds that limiting gun ownership works. Thomas Hamilton, forty-three, unloaded his weapons on sixteen children at Dunblane Primary School, near Stirling, Scotland, in 1996. Nothing like this had ever happened in Europe. The following year, the English government responded to popular pressure by prohibiting the bearing and possession of semi-automatic weapons. Crime committed with firearms has declined significantly since then, and no other killings have been committed in Britain's schools.

In the last few decades, Japan has maintained a strict policy limiting weapon ownership. Their gun crime rates have declined steadily since 1959, after Japan legislated that "no person shall possess a firearm or firearms or a sword or swords." In 2014, six people died from gunshot wounds, and in 2015, only one. Citing police records, the *Guardian* reported in 2016 that only 271,000 people legally owned weapons in Japan, a country with a population of 127 million. The policemen, who go unarmed, train in defensive martial arts. It's important to mention that inequality in Japan is also low, which correlates with low crime rates in general.

The Czech documentary *Teaching War*[41] examines how the militaristic spirit grows in our societies. It also alerts us to the ease with which we allow ourselves to be manipulated by the media, which incites our paranoia against enemies that today can be the Russians or the

---

[40] Ibidem.

[41] https://www.filmcenter.cz/en/czech-films-people/52-teaching-war

Muslims and tomorrow the Chinese or the Koreans. It also warns about the effort some countries are making to restore military service in preparation for a possible next great war.

It would seem that in the West, the terrible cost in loss of human lives caused by numerous civil wars, two world wars, and wars in Vietnam and Korea, has not been enough to cause us to question and reform the warlike mentality still prevalent among the great powers.

Japan, on the other hand, learned its lesson after Hiroshima and Nagasaki and maintains its will to build and preserve peace. After all, it's the only country in the world that has been attacked with nuclear bombs. In Article Nine of its most recent Constitution (curiously drafted under the tutelage of the United States, the country that dropped the bombs on a vulnerable civil population), acts of war by the state are prohibited, and war and the use of force as a method of resolving international conflicts are forsaken. Currently, Japan does not even have a national army.

The right to bear arms and the regulations on gun sales are at the center of an endless controversy in the United States.

"A well regulated Militia, being necessary to the security of a free State, the right of the people to keep and bear Arms, shall not be infringed," says the Second Amendment of the US Constitution, written in 1791. At that time, possessing a weapon might have made sense as a way to guarantee citizens their right to insurrect against an oppressive government. But in those days the available weapons were muskets, and flintlock pistols, not semiautomatic rifles like AR-15s, the weapon used on October 1, 2017, by Stephen Paddock when he opened fire on those attending a concert in Las Vegas, killing fifty-eight people and wounding 851 in only ten minutes.

In 2008 the US Supreme Court ruled that the right contemplated in the Second Amendment did not preclude long-standing prohibitions, such as those limiting possession of firearms by felons and the mentally ill or the use of "dangerous or unusual weapons."

The controversy intensifies every time a massacre occurs in which armed civilians attack unarmed civilians, or when a small child dies in an accident by a firearm carelessly left within reach. A few days before I wrote these words, a twelve-year-old girl almost killed a couple of schoolmates when a semiautomatic pistol she was carrying in her backpack accidentally went off. What was a gun doing in her school bag in the first place? About twenty-five children die every week in the United States because of injuries caused by accidentally discharged firearms.

The militaristic mentality also shows in the solutions devised to curtail problems created by violent behavior. Just after the Parkland massacre, the US government raised the possibility of arming teachers. And barely one month after those deaths, a schoolteacher in California, a former police officer trained in the use of weapons, accidentally fired his gun in a classroom while he was demonstrating how to handle it safely. Three students were injured.

In 2012, Bogotá, Colombia, placed a three-month moratorium on the open carrying of handguns without a license or permit. The successful decrease of violent crime in the capital led to the government expanding the policy to the entire national territory in 2015. The homicide trend in the country is now showing a decline. Other countries that have instituted regulations restricting the purchase and use of weapons have seen declines in deaths caused by firearms. However, in the United States, there are weapons in nearly 43 percent of homes, and the rate of deaths caused by firearms is higher in the US than in any other country (almost forty thousand deaths reported by the CDC in 2017 alone).[42]

The words of General Ludendorff, the villain in the 2017 film *Wonder Woman*, are recorded in my head.

Ludendorff tells the heroine, that war "gives man purpose, meaning, a chance to rise above his petty mortal little self and be courageous. Noble. Better."

Judging by the huge military budgets of the great powers, the persistent strategy of maintaining armed conflicts, and the generalized use of weapons, Ludendorff's statement seems to accurately reflect the philosophy behind the use of military might and the desire to sustain the warmongering vocation of those in power. But does war give man a purpose? What a curious way to understand life. Still, war (and violence) might have a purpose—in a world ruled by greed.

In his famous prophetic novel *1984* (published in 1949), English writer George Orwell conceived a world divided into three large blocks or continents that have an agreement to keep the planet in an everlasting war (any two blocks allied against the third), a strategy that these superpowers have devised to justify a notoriously repressive and intrusive system but also their governments' failures. People have no chance to improve their standard of living and the State has an excuse for the shortcomings of the economy:

> But it was also clear that an all-round increase in wealth threatened the destruction—indeed, in some sense was the

[42] R. Igielnik and A. Brown, "Key Takeaways on American's Views of Gun Ownership," Pew Research Center (June 22, 2017).

destruction—of a hierarchical society. In a world in which everyone worked short hours, had enough to eat, lived in a house with a bathroom and a refrigerator, and possessed a motor-car or even an airplane, the most obvious and perhaps the most important form of inequality would already have disappeared. If it once became general, wealth would confer no distinction. It was possible, no doubt, to imagine a society in which wealth, in the sense of personal possessions and luxuries, should be evenly distributed, while power remained in the hands of a small privileged caste. But in practice such a society could not long remain stable. For if leisure and security were enjoyed by all alike, the great mass of human beings who are normally stupefied by poverty would become literate and would learn to think for themselves; and when once they had done this, they would sooner or later realise that the privileged minority had no function, and they would sweep it away. In the long run, a hierarchical society was only possible on a basis of poverty and ignorance.

The masses, said Orwell, will not dare rebel in times of war without risking being court-martialed for treason. War contributes to the preservation of the structure of a hierarchical society and the privileges of the cast above.

The famous Chinese philosopher Confucius (551–479 BC) argued that, by nature, men helped one another when they were happy and attacked each other when they were angry. He also concluded that war resulted from anger and that it was innate to man. However, from a historical perspective, there seems to be enough evidence that wars are motivated by economic considerations.

The first known organized wars were agrarian (around 4,000 BC), which were endemic conflicts over usable land. During prehistoric times, when communal societies prevailed in the world, no organized states existed, there were no production of surpluses, and no armies were necessary to preserve the status quo. In such circumstances, what we call war was unnecessary.[43] No cave paintings from the upper Paleolithic representing violent confrontations between people have been found. Only one sample of cave paintings (about 10,000 years old) seems to represent a violent scene between humans. An article by Marta Mirazon Lahr discusses the finding of 27 battered human skeletons in Nataruk (now Kenya—possibly also 10,000 years old),

[43] M. Lahr et al., "Inter-Group Violence among Early Holocene Hunter-Gatherers of West Turkana, Kenya," *Nature* 529 (7586) (Jan. 21, 2016): 394–8.

which are believed to belong to nomadic hunter-gatherers, and has been interpreted as evidence of the existence of interpersonal violence in prehistoric times.

How could empathy fully develop in a hostile world? And, really, what is admirable about war? Could we start considering adopting a narrative of love instead?

A military worldview does not contribute to developing skills such as dialoguing, negotiating, offering mutual respect, or building autonomy.

## Which side are you on?

*He who is not angry when there is just cause for anger is immoral. Why? Because anger looks to the good of justice. And if you can live amid injustice without anger, you are immoral as well as unjust.*

—Thomas Aquinas

Most of us are rightly indignant when we learn that there are currently twenty-two countries at war—without counting all of the violent conflicts presently coexisting in the world with an underdetermined number of children serving as soldiers, including 203 reports of children used as suicide bombers. Besides, UNICEF reports two-hundred and forty million children living in areas affected by ongoing conflict.

Such indignation proves we are empathic people, capable of feeling the suffering of others, capable of reacting to social injustice. This discomfort is necessary. It is crucial to experience a certain degree of indignation, empathy, interest, and commitment about the state of affairs on this planet. Otherwise, we watch and react to the news, but the feeling flies away, the terror is forgotten, and no actions are taken to transform the world.

Reporting in the magazine *Foreign Policy*, about the increase of the number of wars in the world, Jean-Marie Guéhenno[44] discusses how we have reached a tipping point at which we are no longer capable of facing the consequences of these conflicts. How are we going to shelter and feed the refugees? How do we grant them access to health services? How do we transport them to safe places? European countries, especially Italy, Germany, and France, have been struggling in the face of the constant

[44] Jean-Marie Guéhenno, "10 Conflicts to Watch in 2017," www.foreignpolicy.org. (January 2017): Retrieved online April 22, 2019.

arrival of immigrants and refugees from Middle Eastern and African countries, mostly Libya. The European Union has already signed agreements with Turkey, Afghanistan, and several African states to reduce the entry of immigrants and refugees, and the future will probably bring new immigration restrictions.

Guéhenno says,

> The world is entering its most dangerous chapter in decades. The sharp uptick in war over recent years is outstripping our ability to cope with the consequences. From the global refugee crisis to the spread of terrorism, our collective failure to resolve conflict is giving birth to new threats and emergencies. Even in peaceful societies, the politics of fear is leading to dangerous polarization and demagoguery.

On the other hand, the proliferation of terrorist groups—the highest expression of human folly—keeps the world in suspense, and it's no surprise that fear often becomes the emotion governing our decisions: from driving to a shopping mall to attending a street festival or concert or going anyplace that has already been a target. Some governments have started to ban entry to natives of countries that finance terrorism or harbor a large number of terrorists.

Our failure to solve the refugee problem is clear. Governments have taken positions ranging from extreme protectionism to remarkable compassion. But for the ordinary citizen like you and me, the feeling is of a great impotence. The problem is of such magnitude that even if we ask ourselves how we could help, there are no straight answers.

I see another form of calamity here. The proliferation of wars, tribal and gang conflicts, terrorism, and even the treaties that force the United States, Russia, or the United Nations to get involved in a war if an ally is attacked, all of this seems to be the antithesis of civilization. There seems to be a misunderstanding as to what it means to rise together as solidary communities, capable of solving our problems through means other than the use of force.

While some studies find a clear correlation between empathic outrage and a display of aggressiveness in defense of a victim, it has also been observed that indignation can lead to the search for necessary solutions. It leads us to take sides and take action.

The young Pakistani Malala Yousafzai is a good example of how a sense of injustice, a righteous indignation, moral anger, could lead to altruistic behavior. She was the victim of a terrorist attack in 2012, when she was only fifteen years old, because she dared to oppose the extremist group that prohibited the education of girls in her country. Malala survived her injuries, and her courage has earned her the sympathy of thousands of people. Her cause rapidly spread through social media. She transformed her sorrow into strength and is now a recognized world activist who fights for the right of girls to get an education. She is the youngest person nominated for the Nobel Peace Prize, which she was awarded in 2014.

Indifference or a noncommittal stance, on the other hand, builds nothing, solves nothing. Individualism often leads to ignoring reality, which is understandable because it can be overwhelming. However, we could contribute in many ways to the needed solutions, and it's for each one of us to find out how.

As the Italian philosopher Antonio Gramsci said, "I believe that living means taking sides. Those who really live cannot help being a citizen and a partisan."

## The immorality of our economy

*Money must be another contender—so many lives are filled with dreams of it, pursuit of it, spending it. It's a faith with many faces: credit cards that let us buy more than we can afford; houses for which we borrow and borrow; lottery tickets that we know make little sense. Perhaps this is Marx's ultimate defeat: is capitalism now the opium of the people?*

—Rosie Blau

Thanks to the advances in telecommunications, transportation, and the internet, North American culture—of which we feel so proud—has infiltrated the whole world without restriction, tainting local customs, traditions, modes of production, consumption, lifestyles, and even ways of thinking. But what has become global is not, as it's sometimes falsely represented, a set of models of social justice and freedom, which foster democracy. Neither is this expanding influence about promoting universal access to science, which would open the way to much-needed solutions for the most pressing problems affecting the world. No. The United States' influence around the planet neither comes from cultural expressions inviting us to appreciate beauty in its most sophisticated forms nor enriches our spirits in the way the Greeks once inspired the West through art and thought or in the way Europe's Enlightenment in the eighteenth century influenced the Americas with ideals of liberty and progress.

Nor is this cultural penetration about free expression of the individual. Instead it's a clever manipulation of the masses for the sake of increased consumerism, the propagation of sectarianism and large-scale group pressure, and the proliferation of McDonalds (thirty-eight thousand stores in 120 countries) and Starbucks (fifteen thousand stores in fifty countries) and other chain stores throughout the world as symbols of neoliberal expansionism.

I consider immoral the deceptive marketing strategies that target and affect people's lifestyles and preferences, that create needs and wants to stimulate consumption of goods, forcing consumers into a spiral of debt and dissatisfaction.

But also, let's consider the consequences for the health of millions. One example is diabetes, today one of the leading causes of mortality in the world. Diabetes has dramatically increased with the spread of Western-style diets in developing nations. Obesity rates are also growing at a rate that matches that of the displacement of traditional diets by junk food and a more sedentary lifestyle.

As a presidential candidate, Bernie Sanders cried out against the suffering of millions of people as the result of the reckless behavior of Wall Street and the lack of ethics of large soulless corporations, whose only purpose seems to be the accumulation of wealth. Sanders asserted that it's time to build what Pope Francis called a moral economy based on fairness, solidarity, and justice.

The behavior of corporations is an expression of egocentricity and selfishness. They are motivated by profit, regardless of the conditions under which people work or whether or not they lose their jobs. Where is the ethics in that? Sanders asks. For him, morality is caring about what happens to others.

Because economics is the driving force of politics, we need to look at the political and social consequences of a global economy. While policies have facilitated cross-border trading and made markets more efficient, it hasn't delivered wealth equality, as some economists predicted and promised. On the contrary, the gap between rich and poor, a direct consequence of the voracity of corporations, has become extreme, especially in developing countries. It has also caused environmental degradation.

The immorality of our economy is also seen in the fact that, "a handful of rich men headed by the Microsoft founder Bill Gates are worth $426bn (£350bn), equivalent to the wealth of 3.6 billion people"—the poorest half of humanity (Oxfam International—www.oxfam.org). Oxfam report in 2018 established that 82 percent of the wealth generated in 2017 in the US went to the richest one percent of the population, while the poorest half saw no increase in their income.

The accumulation of wealth, typical of capitalism, is extreme in the era of globalization. Notwithstanding the economic growth prompted by president Barak Obama's economic stimulus in the United States after the great recession of 2008, this growth actually benefitted only a tiny percentage of the population. And despite Trump's claims of a "booming economy," nominal wage growth still hasn't caught up to pre-recession levels and economists don't seem to understand why.

"When you've got an economy that does great for those with money and isn't doing great for everyone else, that's corruption, pure and simple," Elizabeth Warren has said.

In other countries, the situation is not very different from that of the United States. In Colombia, for example, one percent of the population holds 40 percent of the wealth and it's worse in the countryside. Oxfam reports that according to the 2014 census in Colombia, "1% of the largest landholdings now occupy 81% of productive land and that this extreme inequality has grown worse over the past half century."

In Spain, twenty people have as much money as 30 percent of the population. Reports from the Economic Commission for Latin America and the Caribbean (ECLAC) show that the wealthiest 10 percent of Latin America and the Caribbean holds 71 percent of the riches. Look for figures at the World Inequality Database website.

It is iniquitous that while some people lavishly spend money on nonessential items and we waste goods and foods, around 80 percent of the world's population is still making less than ten dollars a day (minimum wage in Colombia is about US$250 per month, for example).

The biggest immorality is not in being rich or in keeping your surplus money, as the *Current Affairs* journalist A. Q. Smith suggests in one of his articles. "Being extremely wealthy is impossible to justify in a world containing deprivation," the writer said. But, I think, the biggest immorality is in the way wealth is made.

That man's innate drive to promote his self-interest could lead to economic prosperity, was an idea popularized by the Scottish economist and philosopher Adam Smith, who authored Nature and Causes of the Wealth of Nations in the eighteenth century. The free-market economic theory is based on his ideas. And Berkeley researchers Paul Piff and Dacher Keltner actually found that wealthy people tend to think selfishness is a virtue. However, they observed that as people became richer, their compassionate feelings towards other people declined.

But the old question persists: what was first?

Does selfishness actually lead to wealth? Does wealth make you selfish? Or, maybe, these are all wrong questions.

Given the cost to our health, relationships, and the planet, perhaps we should be asking if it's wise to continue to pursue the accumulation of wealth.

## Globalization and the frailty of human bonds

In his book *Liquid Love: On the Frailty of Human Bonds* (Blackwell Publishers, 2003), the late Polish sociologist and philosopher Zygmunt Bauman discussed the frailty of our human bonds in these times of insecurity, typical of a continually changing and globalized world. He said that in modern life, characterized by liquidity (fluidity), "all agreements are temporary, fleeting, and valid only until further notice." There are no permanent bonds, said Bauman, and attachments may hold for a time but should be left loose enough so that they could dissolve as quickly and easily as possible if circumstances change.

Young people start their cohabitating at an early age, but often without formalizing their relationships in front of society. Sometimes this is simply for practical reasons, such as sharing a room when they do not have enough resources to live on their own. Cohabiting without a long-term commitment creates a more relaxed bond than that established by marriage. And even a civil union seems an easier bond to break than that of a religious marriage, which usually comprises a vow of staying together "until death do us part" and a strong belief in the indissolubility of marriage. In the last forty years, the percentage of free unions and civil marriages has increased significantly. In a 2011 press release, the US Census Bureau reported that "among all people 15 and older in 2009, 55 percent had been married once, with 30 percent never having been married at all.... At the same time, 15 percent had married more than once, including 12 percent who had married twice and 3 percent who had married three or more times."

The rural family, typical of the feudal societies of yesteryear, largely disappeared with the spread of capitalism, the formation of urban centers, and the displacement of most of the political and social functions of the family to the state. Today, many young people are separated from their parents when they're just getting out of their teen years when they leave to work or to study. It was common for young people to stay in the country and adopt the trade, business, or profession of the family. One wonders if, as they began to break their roots, these young people didn't lose their sense of belonging, their attachment to their communities. And if, once incorporated into the city, didn't they fall victim to the sensationalist entertainment?

"We're all uprooted and anxious now," The *Guardian's* journalist Stuart Jeffries says, in his April 2003 review of the above cited Bauman's book. "Admittedly there's little in the way of specific class analysis here, for it is Bauman's view that all our traditional bonds are loosening their choke holds. Those purportedly fixed and durable ties of family, class, religion, marriage and perhaps even love...aren't as reliable or as desirable as they were."

With the advent of capitalism, the individual began to survive without depending on the family, and as women became aware of their rights, their sense of a social obligation to marry ceased. Now they could choose a partner of their liking without being outcast.

The word *romantic* has been used for centuries, sometimes associated with aesthetics more than love. In the eighteenth century, fueled by novels that depicted and idealized love affairs, romantic love gained weight as an expression of individual freedom. The conditions became favorable for the restoration of equality of rights between sexes (matriarchy preceded patriarchy), and for women, it meant greater independence and more active participation in social causes.

The downside of the new era is that the solidarity that characterized the working class and the sense of community typical of peasant communities also fell apart.

Recent examples of what happens to agricultural and indigenous communities forced to abandon their traditional lifestyles are seen everywhere.

The Nukak Maku is a small community in Colombia that, once upon a time, inhabited the remote rainforests of El Guaviare, near the Amazon forests. Just a couple of decades ago, they still lived peacefully as hunters and gatherers, but they were uprooted from their natural sites by the FARC guerrillas and by colonists growing cocaine. Today the Nukak are in danger of extinction, like the thirty-four other Colombian indigenous tribes, because of hunger and lack of work opportunities. Their situation was aggravated by the use of alcohol and drugs and the advent of illness brought by their contact with "civilization."

In communities like that of the Nukak, when a way of life ends, traditions wither, and finally people are adrift in a society that does not assimilate them but corrupts every aspect of their lives.

The story of the Nukak is that of many other communities on all continents of this planet. "Progress" comes to them in the form of either threatening weapons or steamrollers and bulldozers. "Civilization" banishes them from their original paradise; invades their territory; pollutes their air, their water, their lands; and destroys the ecosystems that sustain life. Every day their situation grows more critical.

I've already mentioned the desertification of the lands in Africa caused by overgrazing, the misuse of soil and water, the cutting down of trees, open-pit mining, the heavy machinery that compacts the soil, and the burning of wooded areas in the quest for more land for large-scale agribusiness. Desertification has led to the ruin of land, extreme poverty, and famine. But in addition, the ties that bound the communities have been broken, rivalry between tribes has increased, young people have adopted unhealthy lifestyles, and, as if all that were not enough, massive migration to urban centers is taking place.

## Polarization, indifference, lack of compassion: ills of the modern world

Ideologist Zygmunt Bauman endorsed the opposition to the G-20. In his prolific writings (sixty books), he discussed the consequences of constant and relentless change in the modern world. Everything—fashions and interests, dreams and fears—is in constant flux. Our relationships, our identities, and even the global economy change rapidly. There're no solid institutions anymore. But he also believed that kind of fluidity provided the necessary conditions for the blooming of a truly autonomous individual and a truly democratic society.

Taking the side of society's outcasts, Bauman warned about the power of corporations. He noted the growing polarization between the elites and the rest of the population and the general public's disinterest in the fate of thousands of people who live in subhuman conditions. He also warned that trade disputes would affect the lives of the most vulnerable populations and individuals.[45]

Bauman and other opponents of the G-20 have protested the fact that a handful of bureaucrats and heads of state have arrogated the right to make decisions about the fate of millions of human beings and, in addition, have coercively imposed those decisions upon other countries. In short, they oppose the extension of the patriarchal model to global power relations.

The G-20 is a forum in which nineteen wealthy nations and the European Union are represented (out of the existing 195 countries). They claim to have gathered out of concern for the destiny of the world with their sights set on ensuring global access to water and food, preservation

[45] Bauman's core ideas are outlined in Ricardo de Querol, "Social Media Are a Trap," *El País* (January 25, 2016).

of the environment, and the prevention of a nuclear threat. However, the forum's opponents believe its actual motives are not altruistic. On the contrary, critics say, even if they pose as redeemers (and as crusaders for the environment and the planet through the Paris Agreement), the G-20 looks more like a congregation of predators plotting the future distribution of wealth and access to the world's resources. This group doesn't seem to respect other nations' cultural legacies or idiosyncrasies. Dissenters believe the group is anticipating that, given the challenges presented by the consumption of fossil fuels in times of global warming, alternative sources of energy will inevitably become the most profitable businesses of the twenty-first century.

When the nations of the world feel forced to compete for resources and power instead of exercising their sovereign prerogatives to develop ways of life based on self-determination, voluntary association, and peaceful coexistence, no one wins. Not even the most powerful countries on earth win, those who oppose the G-20 say.

And what are the guiding motives of the Group of 20, which pretends to play a redeemer's role?

Let's ask questions, many questions. Be curious. It's important to understand to what and to whom we owe financial crises, global warming, ethnic violence, and refugee and immigration crises. There is significant blaming coming from all directions. In whose hands is power concentrated on this planet? Where must the solutions to our tribulations come from?

I do believe we are the solution, if we together take action to transform this planet into a kinder one.

In an agonizing capitalism where markets are first, greedy corporations compete to the death to control the natural resources, the media, and the government. In this context, the word *democracy* loses meaning, and populism gains ground instead, feeding the illusion that some of these leaders, posing as redeemers, will represent the interests and voice of the people. Demagogues champion the illusion of a kind of freedom and fraternity that do not exist and which they can't provide. Nationalists appeal to an illusionary territorial grandeur. And it's legitimate for us to respond to inequality and famine and exodus with righteous (empathetic) indignation. It's legit for us, the people, to demand that responsible parties repair this disastrous state of affairs in the world. But as individuals, we also have a responsibility and a role to play.

When the right of peoples to self-determination is not respected (disrespecting it is a form of aggression) a germ of indignation grows, generating anger and violent expressions. Disputes over natural resources lead to wars, invasions, and innumerable human tragedies. If peaceful coexistence, collaboration, and voluntary association are not endorsed, there is no fertile soil for love, compassion, solidarity, or even real progress.

Just consider the psychological suffering of the children exposed to the horror of war. The thousands of children recruited to fight on the front lines or forced to commit suicide bombings. Think of the little ones putting their lives in harm's way by working as messengers in conflict zones. The exoduses, the daily increase in the number of refugees, the living conditions in the refugee camps, the scarcity of resources available for those who flee from war should be enough to corroborate that we are definitely not on the path of love. On the contrary, we're moving in the opposite, wrong, direction. And therefore, we can't call it progress.

Advances in scientific knowledge and productive technology in the last hundred years—computer engineering, genetic research, digital communications, brain science, and nuclear medicine—are admirable. However, just as technological and scientific developments could save us, they can also lead to a monumental and irreparable disaster. Just an example: Mashable has recently published an article about the use of cobalt in the lithium batteries of iPhones and electric cars. Cobalt is a rare metal. Respect for human rights in cobalt mines, like those in the Congo, is questionable.[46] Apple is already taking steps to control access to mines and ensure that the 1.4 billion iPhone owners in the world will not be left without replacement batteries. Now, just visualize where all used phones end up when we dispose of them.

The process of globalization has been characterized by the supremacy of the rich countries and the cultural penetration of Western culture—presumed to be superior, democratic, and right—into other countries. National identities have been ravaged and disfigured, and any certainties about the future have been swept away. Fears are propagated by the media, and the world is kept under a constant, not-so-subtle threat of war.

The refugee crisis has contributed to new forms of territorialism. It has generated anxieties that have led many countries to consider tightening their immigration policies. Refugees are mistrusted and resented. The neighbor becomes a suspect. Throughout Europe, populism is gaining ground. The charisma of a demagogue has become more important than his ideas or qualifications to lead.

Natural disasters, economic crises, and dizzying changes in our lifestyles have fueled the kind of fear that rouses us as humans every

[46] Refers to mines that are the source of at least half the cobalt used by companies such as Microsoft, Tesla, and Samsung.

time we face uncertain situations. People's fears are no longer local or manageable. The radical nationalist movements that remind us of the terror of Hitler or the worst days of the KKK seem to be reemerging. Recently, Italy gave the populist right wing a win. A sector that has been characterized as nationalist, xenophobic, and opposed to the European Union obtained significant gains in the May 2019 EU parliament elections.

The interconnectedness of the world might be a blessing to many, but it's also perceived as a curse. Let's remember that all countries having direct trade links with the United States bore the impact of the recession of 2008.

As mentioned before, there's an interesting paradox in that, although globalization has brought interconnectedness to the markets and people's destinies, it has also made us increasingly aware that we are just one people inhabiting the same solitary planet.

Because of this interconnectedness, it's essential that we support each other, that we accept unity in diversity. It's also crucial to understand that whatever measures we adopt and whatever crises we endure "here," will have repercussions "there" (or better said, there is really no "here" or "there").

Given the fiasco of the autocratic bureaucracies of the so-called communist regimes, one could give in to the illusion that the capitalist system has succeeded like no other economic system. We can't ignore the achievements of an impressive technological development or the fact that many health, communication, and transportation challenges of the early twentieth century now seem to have been overcome. No doubt our lives are ever more comfortable, and domestic chores have become a breeze. New farming techniques and livestock breeding, plus the use of fertilizers, herbicides, and irrigation systems have increased farming productivity. Digital communications are amazing. The field of assisted reproductive technology is rapidly progressing. Knees and hips are easily replaced. We can even transplant whole faces and communicate via satellite. We are every day closer to traveling to Mars.

But how can anyone claim the capitalist system has been successful when many of the inhabitants of the Third World still fall victim to famines, ethnic conflicts, and mass exoduses, and when this seems to be the outcome of precisely the same greedy practices of corporations that have made possible such great advances? A system where so many talented people still lack job opportunities and almost 1.4 billion workers were estimated to be in vulnerable employment situations in 2017. Where children and women fall victims of human trafficking? Where every 11 minutes someone in the US dies from an opioid overdose? How can we talk about the success of a system that glorifies junk food while our lifestyle choices increase the chance of suffering from cancer, diabetes, or coronary disease?

Many people whose livelihoods have become unsustainable have lost hope of a better future and have fled their homelands, hoping to take refuge in a rich, industrialized country seen as a promised land. They follow their dream and sometimes end up living in dismal conditions, significantly isolated for cultural reasons and language barriers.

In my view, an economy can't be called successful if it can't solve the problems mentioned above for the majority of the people, if, in the best of the cases, it leaves people only moderately satisfied, if it offers working conditions and wages that fail to fulfill the material, emotional, and spiritual needs of the individual.

It's difficult to call a socioeconomic system fair when 40 percent of the food produced in a country like the US is thrown away (not counting the food that rots in our refrigerators), while not very far away children are dying of hunger, and the law prescribes that supermarkets and restaurants throw away the cooked food left at the end of the day instead of using it to feed the needy.

Disturbingly, according to statistics published by the Food Aid Foundation, at least 795 million people in the world do not enjoy food security even though the amount of food produced in the planet is enough to feed everyone in the planet. According to the definition of the United Nations, food security is "the condition in which all people, at all times, have physical, social and economic access to sufficient safe and nutritious food that meets their dietary needs and food preferences for an active and healthy life."

The way in which wealth is progressively concentrated in a handful of people goes rather unnoticed, while the middle class in Europe and North America become impoverished and Africa dies of hunger.

Yemen, South Sudan, Somalia, Ethiopia, Chad, and Sudan were the countries with the highest rates of hunger in 2017, according to the Global Hunger Index (which doesn't calculate hunger in the Unites States), taking into account three indicators: nutrition, weight, and infant mortality. On this side of the Atlantic, Haiti is going through circumstances similar to those of agricultural countries in Africa. Part of the problem is not even an actual shortage of food, but the lack of infrastructure required to transport it to where it's needed. On the other hand, the growing deforestation, the monocultures, and the effects of climate change have affected the availability of farming land. In Haiti, 80 percent of the population lives below the poverty level. In the countryside, only 10 percent of the population has electricity generated by fuel oil and the rest uses charcoal as fuel, which causes serious problems of desertification, erosion, and landslides.[47]

It's increasingly evident that the greed of the corporations is at the root of the humanitarian crises across the planet. One North American corporation, Cargill, tripled its profits in just one quarter in 2010 by speculating on the stock market with essential agricultural products. And the trend is that the more food produced, the smaller the number of companies marketing it. As they grow huge, corporations can dictate prices, maintain low wages, determine the terms and conditions of trade, and press for political changes that favor them. African farmers, for example, are pressured to grow what corporations pay at better prices: coffee, cotton, and grain that will be consumed by rich countries.

Econexus[48] offers an example of a corporation marketing a particular type of fish (pangasius), which is produced and exported by the Vietnamese. Northern countries pay $10 a kilo for it. The person who grows the fish receives $1 a kilo. After deducting costs, he's left with 10 cents, and this is without considering production risks.

Stephen Leahy reports[49] how in a corrupt global food system, farmland has become the "new gold" and Africans the "new share-croppers." Recently, investors from Saudi Arabia leased nearly a hundred thousand hectares in Ethiopia, Senegal, and Mali for ninety-nine years.

The accumulation of wealth continues to grow. Injustice and social inequality do not breed love or compassion, and neither do ambition or greed. Instead they promote competition, power struggles, fear, and envy. Social injustice dehumanizes us all.

## Debilitating stress

In my book *Regaining Body Wisdom (2008)*, I discuss stress as a constant of life, explaining how our capacity to respond to stressors changes through our life cycle. When we think about stress, we more

[47] Deforestation in Haiti is up to 98 percent (the worst deforestation in the world). Proceedings of the National Academy of Sciences (PNAS), has suggested that if action is not taken, the country will lose all of its remaining forests over the next two decades. Is it fair to blame the people for the cutting of the trees? Since colonial times, the forests were cleared to grow sugar cane and coffee and also to export the wood.

[48] Visit: http://www.econexus.info

[49] S. Leahy, "In Corrupt Global Food System, Farmland Is the New Gold," *Intrerpress Service* (Jan. 2011).

often focus on one form of it: psychosocial stress, brought about by social situations and environments in which we feel threatened or strained. However, stress refers to any environmental changes or demands perceived by the physical body, the mind, or our energy field, independent of the stressors' nature.

Fortunately, the body is well-equipped for behavioral, physiological, or structural adaptation to increase its chances of survival. Stress is part of a normal process of interaction between human beings and their environment; it's part of life. What is new is chronic and cumulative stress, which is a characteristic of modern life. The level of stress has risen significantly since the industrial revolution, but never before the digital era were human beings subjected to such a variety of stressors, and with so few restoring pauses or compensatory factors.

Our multidimensional body is frequently exposed to various stressors: psychosocial, physical, mental, spiritual, biological, chemical, or electromagnetic. All stimuli that impose unusual demands on the body—including acute exposure to extremely low-frequency electromagnetic fields (ELF-EMF) from appliances and medical devices that have the potential to alter our natural frequencies and affect our nervous system—cause stress.

If stressors are intense and/or constant, they can cause discomfort, pain, or damage to tissues. The detrimental effects of stress on the body are explained by the increase in free radicals—highly reactive ions that cause oxidative damage to cells. When our lifestyle doesn't offer the ideal conditions to respond to them, stressors can lead to metabolic disturbance and symptoms of physical, emotional, or mental disease.

Precisely because stress is a common denominator to our ailments, we have come to learn quite a bit not only about those biological, physical, mental, emotional, energetic, and spiritual factors that generate stress, but also about the ways in which the multidimensional body responds to those adaptation demands from the environment. The most common stressors in modern life are chemical and psychosocial. Cavemen feared the tiger; we fear terrorist attacks and war, the catastrophes associated with climate change, the chance of losing our livelihood. Noise, traffic, visual and auditory pollution, water and air pollution, isolation, highly competitive jobs, and difficulties in interpersonal relationships are also stressors typical of our times.

Wilkinson, mentioned above, discusses the psychosocial effects of inequity. The greater the inequality, the greater the insecurity people experience. The sociologist cites psychosocial studies showing that cortisol (a marker for stress) is higher in people living in conditions of social inequality, where you fear being judged socially, where you constantly compete with others (who is the most intelligent, beautiful, rich), and where your performance is under steady scrutiny. Interestingly, feelings of inferiority or superiority, our need to be valued and respected, could stimulate consumption of goods that provide an appearance of status. In that sense, inequity turns out to be good for business.

We are particularly interested in stressors typical of our times and how every stressful psychosocial event might elicit fear to which we usually respond automatically. We might withdraw from social life, isolate ourselves, confront or compete with others, or try to make friends and seek support. Each option is associated with the activation of distinctive emotional responses in the brain, some of which are not favorable to the development of empathy.

Please note that there is no such thing as "negative emotions." Emotions help us navigate the environment by alerting us to the existence of threats and rewards. However, as a tool, and due to previous experiences, emotions might lead to less than optimal, automatic responses. Therefore, it's crucial that we learn to consciously modulate the type and intensity of the emotions we feel, when we feel them, and how we express them.

Stressful situations have the potential to either unite or separate us—revitalize or exhaust us. Stress induces a set of powerful behavioral, emotional, hormonal, cellular, and molecular responses that mediate the process of adaptation of an organism to the challenges of the physical or social environment. The effects of chronic stress on the multidimensional body are often detrimental.

The survival of a relationship, the closeness between children and parents, getting along with neighbors, taking leadership, become significantly affected by the stress levels and how we handle them. High levels of stress interfere negatively with any dimension of our lives. The good news is we certainly have options; we can train our responses to stress to make sure we strengthen, instead of crack, our bonds. More on this later.

## Stories that cloud our perspective

We build or adopt a series of explanatory models or beliefs that shape the story of who we are, using this story to try to elucidate the world we live in. The circumstances in which we grew up, the way we were raised, our family's composition, our community and school, the impact of personal experiences, the media, are all factors that influence how we see the world. At least some of our perceptions might contain cognitive distortions or false assumptions that limit our potential. They can affect our behavior, our feelings, our capacity to connect to others, and our learning processes.

However, we cling to these subconscious forces that drive our behavior, even if they compromise the essence of who we truly are. We identify with these tacit stories about ourselves and the world. And even when we gain insight into these beliefs, we find it difficult to replace them because it feels as if our identity—based on those inaccurate premises—will crumble.

Those implicit beliefs could be about control, competency, or self-value. Some of those stories are generalizations:

- I need love (if I don't have a partner, I'm a failure).
- I need approval (nobody understands me, nobody helps me, nobody loves me).
- I broke this, I hurt others, I can't do it (I made the wrong decision, this had to be perfect).
- Something is missing (I'm not good enough, I don't deserve this, I'm not capable, nobody is good enough, nobody can help).
- Something bad is going to happen (if I open up, people could hurt me, I don't feel safe, I have bad luck, I can't trust myself or others).
- All men are or all women are… (and other similar beliefs that lead to chauvinism, racism, nationalism).
- There is only one right way of doing this (I can't/won't change, others are wrong, I'm perfect).

Many of these beliefs are based on stories of separation (us versus them or us versus the world) absorbed from a hierarchical, polarized, and harshly dominant culture. However, these stories don't necessarily reflect our direct experiences. Instead they echo particular relational patterns and the way our society understands power. Good examples are the stereotypes about the roles of men and women or the nature of the relationship between parents and children, which mainly come from an ideology we neither constructed by ourselves nor adopted consciously.

Educational systems are greatly responsible for the transmission of these beliefs. Hollywood also bears a large share of the responsibility, contributing to the perpetuation of stereotypes, ideologies, and illusions. And the news media, instead of prioritizing their duty to inform us, too often prey on our credulity. Instead of ethically separating news from the influence of media owners and advertisers, many writers become experts in journalistic deception due to organizational pressures. Here's an example: In 2015, the former chief political commentator for the *Daily*

*Telegraph* revealed that the paper had been suppressing, removing or discouraging certain stories to avoid offending (or, alternatively, to please) advertisers or financial institutions.

Neuroscience explains that adverse events (referring to thoughts, emotions, social interactions, incidents) have a higher impact on our psychological processes and behaviors than positive ones. We tend to focus, learn, and select negative information more frequently. Our memory is also selective. Of ten incidents happening in one morning, nine pleasant, one unpleasant, we'll first remember the fool who dangerously cut us off while driving, and we'll use the negative event to unconsciously confirm the belief that, for example, people in our city drive recklessly and unkindly. Our mental and emotional well-being comes from what is known and predictable, from what we believe works, at least for us.

Most of our limiting beliefs stem from a negativity bias: we focus on the harmful, on the potentially dangerous. And we need to. The brain evolved to be vigilant and fearful for a reason. To automatically react defensively in front of a perceived threat is economical for the mind and for the body. However, although this bias evolved to keep us out of harm's way, it doesn't mean we have to live an automatic, reactive life.

Paul Rozin and Edward B. Royzman, of the University of Pennsylvania, were the first to hypothesize the existence of this general negativity bias, which is also "contagious." They believed it was based on innate factors and on experience and that it occurred both in animals and humans. They proposed four elements to explain it: a) *negativity potency*: the negative has more strength than the positive; b) *steeper negative gradients*: the negativity of adverse events grows faster; c) *negativity dominance*: the combination of positive and negative events leads to a more negative evaluation of a situation than what could be predicted mathematically; and d) *negative differentiation*: negative events are more varied and lead to more complex conceptual representations, and therefore their responses come from a wider repertoire of experience. [50]

Our negativity bias could become the most significant obstacle to, for example, developing romantic relationships. How do we get out of that conditioning?

During a workshop on neuroscience and expressive writing I attended a few years ago in New York, facilitator Kathleen Adams

---

[50] P. Rozin and E. B. Royzman, "Negativity Bias, Negativity Dominance, and Contagion," *Sage Journals* (Nov. 2001): doi.org/10.1207/S15327957PSPR0504_2

explained how the simple habit of keeping a diary could help us overcome this negativity.[51] Writing calms us down, explained the psychotherapist. It helps us to organize our thoughts through language, which requires the use of neural circuits of the cerebral cortex that can override instinctual reactions.

Writing has a healing effect on the person, irrespective of what they write about. There is no need to write about a traumatic event for the exercise of writing to contribute to a better emotional regulation. And we could always cultivate an optimistic bias by recreating ourselves in a positive way.

People with certain mental health symptoms are often trapped in vicious cycles of negative thoughts (rumination) that interfere with effective emotional regulation. The binomial of negative perceptions and beliefs can become an impediment when there's a need for cooperation or asking for support. The process of coming to an agreement, or negotiating terms, can be complicated by our unconscious holding onto negative stories about ourselves or the world, stories that we've either adopted or elaborated over time.

Certain beliefs may become limitations, preventing us from being open to novel or alternative ideas and solutions, new ways of functioning or relating to other people. Beliefs may even interfere with the adoption of a healthy lifestyle, like when a person affirms that her obesity is "a family thing" to justify her difficulties in keeping her weight down or following a healthier diet.

Who are you? If I answered that I am a woman, a mother, a teacher or a doctor, I'm not answering the question. If we identify with the roles we play, with our professions or social status, we won't be able to build self-esteem on firm ground. We will come to depend on the evaluation that others make of us—or on our achievements, performance, and possessions—to feel valuable. When negative stories we tell ourselves become part of our identity, learning something new, giving in when a logical argument proves us wrong, accepting those who are different, cooperating with others, or just being in a relationship without feeling we're losing ourselves becomes truly difficult.

We all need to adopt our own set of guiding principles and standards, but it's also healthy to eliminate the acquired beliefs that have become limiting. It's one thing, for example, to embrace an ideology that provides a frame of reference to understand the world; it's something very different to become a radical militant that hates those who don't

---

[51] Adams is co-editor, with Kate Thompson, of *Expressive Writing: Counseling and Healthcare*. (London: Rowman and Littlefield, 2015)

agree with them. When it comes to a you-versus-me dilemma, I will most likely pick myself.

The sad thing, the real problem, is not being aware that most of our beliefs have been absorbed from our environment or transmitted to us, not chosen by us. Therefore, we haven't built our worldview selectively. Not everything we observe as a norm or accept as a belief is right, valid, or positive. A heightened awareness will give us the ability to jump out of that train in which we found ourselves mostly by accident. It leads us to be critical of beliefs that have become limiting, that divide us, that lead us to conflict or unhappiness.

## Me first

*The modern Narcissus does not refer to a glorious being, but to a self who is immersed in a fantasy of omnipotence and self-sufficiency, based on a fundamental impotence.*

*In fact, when liberalism transforms subjectivity, it causes a fragmentation of the social fabric where the bonds between men weaken and isolation gives rise to an impoverished self. The modern phenomenon of narcissism is an invitation to reflect on the importance of social ties and establish policies aimed at a disunited community.*

—Florianne Gani

A wave of narcissism—individual and collective—has invaded the United States, and, insofar as its culture and marketing penetrate the rest of the world, this narcissism also contaminates other countries, most notably since the 1960s. It's a wave characterized both by a generalized national chauvinism ("We're the most powerful country in the world." "Out Western culture is superior." "There is nothing like our democracy." "Our evils are the alien's fault.") and the individual's need to obtain recognition. A narcissistic person tends to focus on his own goals, disregarding the common good or the possibility that he might be hurting others.

I use the words narcissist and narcissism in the most popular of their meanings, as an exaggerated love for oneself that impedes the appreciation of others, and not as a diagnosis of a psychiatric disorder.

From an epistemological point of view, narcissism originates in an egocentric vision in which it's difficult to see the world from a perspective different from one's own. Self-centeredness is typical of the way young children perceive the world, but in a healthy person, perception evolves as the brain's prefrontal cortex matures.

The terms *collective narcissism* and *group narcissism* were used respectively by psychoanalysts Sigmund Freud (*Group Psychology and the Analysis of the Ego.* London: The International psychoanalytical press, 1922) and Erich Fromm (*The Anatomy of Human Destructiveness*, New York: Holt, Rinehart and Winston, 1973), to describe a phenomenon according to which the members of a group acquire a magnified perception of themselves, a vision that requires external validation. The rise of leaders like Vladimir Putin in Russia and Tayyip Erdogan in Turkey, and then in 2016, the election of Donald Trump in the United States, all seem related to the emergence of narcissistic personalities who extend their grandiose view of themselves to the countries they govern, generating a new kind of nationalism that appeals to the emotions and pride of certain (narcissistic?) collectivities and undermines other nations and peoples. Take a look at the campaign slogans of these men. Erdogan: "Bring Strength to Turkey." Trump: "Let's Make America Great Again." Putin: "Strong President; Strong Russia."

The technological development and growing popularity of social networks like Facebook changed the way we use our free time and communicate. In the last quarter of 2018, across the world 2.3 billion people actively used Facebook. Internet addiction is a new area of study in psychology, and research shows that the frequent use of social media is strongly related to narcissistic behaviors and low self-esteem.

Olivia Reme, a graduate student at the University of Cambridge, provides the following explanation: "As the social fabric deteriorated, it became much harder to meet the basic need for meaningful connection. The question moved from what is best for other people and the family to what is best for me."[52]

Researchers at the University of Michigan created the Egocentricity Index and used it to compare the annual speeches of American presidents from 1790 to 2012. They measured the number of words that indicated a personal interest compared to the number of words that indicated an interest in others. They found that after 1920 every presidential speech has progressively included more self-interest words.[53]

Apparently, the more prosperous a nation, the more self-centered it becomes.

In another study, University of Emory business professor Emily Bianchi concluded that young people who grew up in times of recession

---

[52] O. Remes, "Why Are We Becoming So Narcissistic? Here Is the Science," *The Conversation* (March 2016): https://bit.ly/2K4Ibe4.

[53] D. Swanbrow, "State of the Nation's Egotism: On the Rise for a Century," *Michigan News*, (March 2014) https://bit.ly/2K9K4Xe.

became less narcissistic than those who had grown up in times of abundance.[54] Bianchi correlates the results of her study to the qualities that a thriving Western society values in a person: autonomy, self-sufficiency, self-confidence, and the ability to advocate for themselves.

In collective narcissism, the individual usually sees his group as an extension of himself and expects others to recognize not only his greatness but also the prominence of his group. In both collective and individual narcissism, egocentrism distorts the perception of reality to the extent that the individual uses his or her feelings and thoughts as a parameter with which to evaluate what the other thinks or feels. We assume the other is (or should be) like us or that we don't need to explain ourselves; the other should know what we're thinking or feeling.

People who are not narcissists assess their abilities and strengths with no need to compare themselves with others but by measuring their own growth and development. If I praise and magnify myself through putting others down, I only create distance and separation, never connection.

Although psychologist Jean Piaget's idea has been debated, it's commonly accepted that human beings go through an egocentric stage as a normal part of their cognitive development (Piaget, J. *Origins of intelligence in the child.* London: Routledge & Kegan Paul, 1936). For example, at first, infants and toddlers don't speak in order to communicate (they don't know the social function of language yet), but to obtain what they need or want. They will not acquire the ability to understand the perspective of others until the age of seven or eight (maybe earlier these days). As they grow, they must develop this ability, a requirement without which empathy will not evolve.

A substantial aspect of narcissism these days is a preoccupation with appearance and financial success. At San Diego State University, Professor Jean Twenge, a researcher of self-esteem in youngsters, has teamed with Keith Campbell, who specializes in narcissism, to discern if youngsters are more narcissistic these days.

Twenge[55] explains how self-esteem is different from narcissism: "Somebody high in self-esteem values individual achievement, but they also value their relationships and caring for others.... Narcissists are missing that piece about valuing, caring and their relationships, so they tend to lack empathy, they have poor relationship skills. That's one of

---

[54] E. Bianchi, "Entering Adulthood in a Recession Tempers Later Narcissism." *Psychological Science,* Vol. 25(7) 1429–1437 (2014).

[55] L. Malcom, "Research Says Young People Today Are More Narcissistic Than Ever," *Australia National Radio*, (May 2014) https://ab.co/QP7Trg.

the biggest differences, those communal and caring traits tend to be high in most people with self-esteem but not among those who are high in narcissism."

Twenge studied symptoms that correlate with narcissistic traits and behaviors and found that the number of plastic surgeries catering to vanity has significantly spiked since the 1990s in the United States. Moreover, in a survey conducted by the same professor, it was found that, among university applicants, 82 percent felt that making money was their most important goal (materialism), while only 45 percent believed this in the 1960s and 1970s. [56]

Narcissism has become a trend. The topic has been trending in newspapers, research papers, and psychology articles. I just Googled the word and got 137,300,000 entries.

Articles published on narcissism are common in other countries. In Spain, for example, they often correlate with corruption in the government. French authors Marcel Gauchet, Gilles Lipovetsky, and Jean-Pierre Lebrun have also addressed the topic and analyzed the characteristics of this emerging self-centered man from philosophical, sociological, and psychoanalytical points of view, trying to understand this new way of existing in the world.

Journalist Reynaldo Spitaletta wrote about narcissism for his column in the Colombian newspaper *El Espectador*:

> Nowadays, the subject ego's center has moved to the body, which can be shaped in gyms with the criterion of superficial hedonism, appearance, presumed physical health. It is not so much the matter of a healthy mind in a healthy body, but of a body without a mind, without limits, mostly out there for display. A love of his forms, muscles, gestures. The key thing now is 'my physical profile', my style, my spending habits, showing off.

Ideological values planted by a society determine in great part if we develop narcissism or healthy self-esteem. The last, entailing a clear awareness of our worth, goes along with a humble attitude toward others and toward life.

Twenge believes the narcissistic self-esteem occurring in the West does not lead to improved academic or professional performance, does not guarantee leadership skills, and does not prevent risky behavior in young people.

---

[56] Statistics by the International Society of Aesthetic Plastic Surgery (ISAPS) published in 2017, show that USA, Brazil, Japan, Mexico and Italy report the highest number of plastic surgery procedures, followed by Germany, Colombia, and Thailand.

In *The Many Faces of Narcissism,* Gabbard and Crisp-Han[57] said, "Members of the millennial generation live in a constantly connected, technologically visible and self-oriented public space. Especially millennials seemed fixed on using social networking to brand themselves. *Time* captured this cultural moment by referring to the Me, Me, Me generation."

Smart phones have given rise to a generation that has lost the pleasure and art of conversation. I've found many of them even have trouble talking on the phone but feel at ease texting, sending memes or emojis, sharing videos, or "YouTubing."

According to author Sherry Turkle, a new personality is also emerging, one shaped according to what others want to see. The validation of the self comes from flattery and self-esteem and swells each time a selfie is liked or receives a comment.[58] In short, the newer generations (millennials, generation Z) are blamed for their growing narcissism, but young people should not be deemed responsible for this world of selfies. They too are victims caught in a consumerist society that promotes self-indulgence and the acquisition of more and more stuff and provides an excess of distractions and illusions. Today's world is full of industries that market and demand images of perfection. According to Orbis Research, the global beauty market reached about $532 billion in 2017.

Our young people grow up in a cheerleading culture, showered with empty praise from an early age, surrounded by adults who send the message that they must become rich and famous if they aspire to any recognition and appreciation.

Fun fact: mathematician Samuel Arbesman came up with a number: only 0.0086 percent of the global population is what we could call famous.

Researchers have found that one characteristic of this new personality associated with the advent of social media and greater wealth is altruism.

Gabbard and Crisp-Han said they regret that narcissism and altruism are often considered antagonists. The two entities coexist, they said. The authors cite researcher George Vaillant, MD, who found that altruism increases significantly after age fifty, not simply because we become increasingly selfless with age, but because helping others

---

[57] G. O. Gabbard and H. Crisp-Han, "The Many Faces of Narcissism," *World Psychiatry* 15(2) (Jun. 5, 2016, online): 115–116.

[58] Sherry Turkle, *Alone Together: Why We Expect More from Technology and Less from Each Other* (New York: Basic Books, 2011).

becomes more rewarding. They also cited a longitudinal study in which Jorge Moll et al. scanned nineteen subjects with fMRI while they anonymously decided whether or not they wanted to support societal causes by donating to real charitable organizations. Participants were entitled to receive US $128, if they solely cared about their interests when making decisions. Ratings of experienced compassion were higher for participants who chose to donate, whereas anger scores were higher for those who opposed causes. Interestingly, the mesolimbic dopaminergic circuits in their brains were activated as much when they donated as when they received monetary rewards.[59]

Many young people today get involved in service projects, participate in politics for altruistic reasons, and on average contribute to charities in greater proportion than their elders. Some argue they do so for egotistical reasons. I see adolescence as an incredibly lucid stage.

Many bullies show a certain pride, which others might consider high self-esteem. However, instead of loving themselves, bullies seem more often driven by shame and a need for control. They often display narcissistic traits, with an idealized image of themselves that they admire and want others to see. Bullies are full of prejudices about others and need to continually demonstrate superiority, which prevents them from creating healthy and long-lasting bonds. They're full of fear of failing or being weak and focus on a pernicious need to be considered the best, the strongest, the prettiest. Another narcissistic trait typical of bullies is a constant fear of being betrayed. That's why loyalty from their pals matters so much to them.

Deep down, the bully is a narcissist, insecure, and afraid of exposing his inadequacies. More on this subject on Part Six of this book.

## I'm okay, you're wrong

In an interview with the *New York Times*, the African American sociologist and university professor George Yancey, an evangelist, says he's found problems on the street because of his skin color and issues in the academic world because he's a Christian. This illustrates the fact that discrimination has different forms and is not exclusive to a particular political party, religion, skin color, gender identity, or social status.

[59] J. Moll, F. Krueger, R. Zahn, M. Pardini, R. Oliveira-Souza, and J. Grafman, "Human Fronto-Mesolimbic Networks Guide Decisions about Charitable Donation," *PNAS* 103 (42) (October 17, 2006): 15623–15628, *doi.org/10.1073/pnas.0604475103.*

Discrimination has many forms, including an institutional bias. A single example: although African Americans represent 12 percent of regular drug users, arrests for drug possession among blacks account for 32 percent of the total drug-related arrests, and their sentences tend to be harsher. The color of your skin could immediately make you a suspect or lead to your exoneration. The prevalence of racism in the United States is acute and getting worse. Lee and colleagues[60] found that 63.10 percent of minorities experience racial discrimination, compared to 29.61 percent of whites.

The divisive style of the 2016 presidential campaign in the US and subsequent mandate and rallies of the forty-fifth president of the United States generated a wave of xenophobic rhetoric and a regrouping of the extreme right. Some attribute the subsequent divisiveness and hatemongering to the president's inflammatory speeches and tweets, and others to a rebound effect resulting from eight years of having a black person commanding the country from the Oval Office.

The words of Martin Luther King Jr., champion of civil rights in the sixties, still resonate: "Love is the only force capable of transforming the enemy into a friend." But, do we have the necessary will to transform the enemy into a friend? Or at least to understand that a person with different ideas is not necessarily an enemy?

Some studies support the idea that coexisting with people who think and act differently leads to more tolerant attitudes, but then one wonders how we can explain the resurgence of racism, hatred, and intolerance in a globalized, diverse world.

This evident discrimination is not just racial and ethnic. There is also discrimination against people from the LGBTQ community, against women, against members of certain religions. For example, one poll found that nearly 35 percent of Americans supported Trump's intentions to prevent the entry of Muslim immigrants, failing to understand that the Islam (the religion of Muslims) and the Islamic State (ISIS—a political entity) are not one and the same thing.

We used to believe that only a certain kind of fanatic discriminated against others based on stereotypes he built, but today we know we all have some bias in our perception of others. However, most of the time we're not aware of the stereotypes we've built for a biological reason: our brain's need to categorize our experience with other individuals and develop conditioned responses from former experiences that will

[60] R. T. Lee, A. D. Perez, C. M. Boykin, and R. Mendoza-Denton, "On the Prevalence of Racial Discrimination in the United States, *PLOS* (Jan. 10, 2019): doi.org/10.1371/journal.pone.0210698.

allow us to protect ourselves, and sometimes this needs to be done quickly.

When the world is rigidly conceived as a hierarchical system, it's really complicated to put oneself in the place of the other, to understand the point of view of the other, and to be moved by his or her suffering. Intolerance, racism, and discrimination are more common in a system that moves on such a vertical plane, where the main feature is competitiveness.

A psychologically rigid person will have more problems adapting to new situations, higher resistance to change, and less ability to regulate emotions such as fear and anger. Consequently, this person will find it more difficult to cooperate and associate with others. Studies in psychobiology also demonstrate that people with rigid cognitive structures find more problems negotiating and reaching agreements. In my practice, I have observed that people who display abusive behaviors have trouble understanding the extent of the emotional impact on their partners of what they say or do, but they seem aware of the controlling potential of their behavior.

A psychologically flexible person will have a greater capacity to consciously evaluate the present moment and, according to what the situation allows, modify their behavior or course of action to adapt to circumstances. We are born flexible. Rigidity traits are learned. The degree to which a person is psychologically flexible could be evaluated with a questionnaire ("Acceptance & Action Questionnaire," or "AAQII").

Several studies seem to confirm a correlation between cognitive processes and political and ideological positions, fields where tolerance and arrogance are at play. In 1950 the Austrian psychologist Else Frenkel-Brunswik had already argued that a person with an absolutist mind had a bias in his perception of the world.

On the other hand, people with a flexible mind adapt easier to the demands of the moment and are capable of changing their repertoire of behaviors based on their sociability. Such people would find it easy to maintain a balance between different aspects of their life and would usually be more open to exploration and experimentation.

A lack of mental flexibility predicts anxiety, depression, poor performance, and inability to learn new things. Authoritarianism feeds on cognitive rigidity. Authoritarian people see the world in absolute terms (what we commonly call black-and-white thinking).

A study conducted at Emory University by psychologist Drew Westen and his team sought to determine the neurological basis of the political decision-making process. Using magnetic resonance imaging, they found a significant difference between the neural processes of individuals who considered themselves centrists or independents and those who were clearly affiliated with a political position. In the latter, the reading of names of different candidates activated neural circuits that process emotions, while in the former, the cerebral cortex, generally associated with rational mental processes, was more active[61].

According to Arthur Brooks, a business professor at the University of Syracuse, NY, the number of politically radicalized brains is increasing, and we're becoming more inclined to hate our opponents.

Believing one's own position is the only correct one is a sign not only of rigidity but also of arrogance. It prevents us from learning from others but also obstructs the appropriation of two important prerogatives: own responsibility in any situation we get involved in, and the fact that, if we lead a conscious life, how we feel, how we act, and how we interpret our experiences is a choice.

Flexibility is a distinctive quality of a person with good mental health. Cognitive rigidity makes it hard to understand other people's points of view or accept change.

If I'm not capable of seeing my responsibility in a relationship or event, I can't change my circumstances or the way I respond. If I feel I can't choose my answers and don't see my options, it will be easy for me to fall into a victim's position.

Another cognitive bias, known as the empathy gap, formulated by economist George Loewenstein[62] a couple of decades ago, proposes that our ability to understand others depends on our mental state (either cold or cerebral or hot or visceral/emotional). An angry person won't understand why the person in front of him keeps calm. A person who's not in love has a hard time understanding how someone who's in love feels.

It's difficult for us to predict how we'll feel tomorrow, when we'll be in a different state of mind. In the same way, it's difficult to remember, once we've calmed down, how we felt when we were angry or terrified.

If you feel the suffering of another in your guts, if it brings tears to your eyes, you are experiencing emotional empathy. Cognitive empathy, more common among health professionals, priests or therapists, refers to being able to take the perspective of another to understand what they feel and think. Compassion is empathy that compels you to take action. According to Professor Emeritus Paul Ekman, at University of California

[61] D. Westen, *The Political Brain: The Role of Emotion in Deciding the Fate of the Nation*, (London: Perseus Books Group, 2007).

[62] G. Loewenstein, "Hot-cold empathy gaps and medical decision-making." *Health Psychology*, 24(Suppl. 4), S49-S56. 2005.

San Francisco, therapists experience empathic resonance in the process of trying to understand what the clients go through and should be on the alert for the experience of two types of emotional resonance: identical resonance (when you realize that someone else is in pain and you feel the pain yourself mostly because you have gone through the same type of experiences and maybe you haven't processed them). If made conscious, reactive resonance might follow (when after feeling the pain you also feel inclined to help).

Our capacity to resonate with others emotionally and cognitively is crucial in healthcare and business. It determines the quality of our relationships, enabling us to share experiences and needs. Sadly, studies show empathy declines during medical training, and patients treated without empathy are less likely to comply with treatment recommendations. Learning to develop cognitive empathy is also important in our relationships with people who are simply different from us in ethnicity, religion, gender identity, or physical makeup.

Anthropologists consciously use themselves as a reference to try to recognize the forces of socialization that operate in a culture. This is known as reflexivity. It involves putting yourself in the position of another individual and trying to see things from their perspective. It's an exercise that promotes cognitive empathy.

Despite growing sectarianism and polarization, there are hopeful signs. Studies show that the world has also become more tolerant in the past few decades. A good example is the greater inclusion of women in society, even in radically patriarchal countries.

A 2017 Gallup poll showed that the number of Protestants in the United States who don't identify with a specific denomination doubled between 2000 and 2016. This could mean people are avoiding hierarchy and structure or that people are looking for churches that don't discriminate against other faiths or communities.

There is also a greater educational inclusion of children with special needs in schools and social life. The National Center for Education Statistics reports that by the fall of 2014, almost 95 percent of six-to-twenty-one-year-old students with disabilities attended regular schools. And, according to a survey by Jean Twenge[63] covering a forty-year span (1972–2012), Americans have become increasingly tolerant even of controversial outgroups: LGBT, Communists, anti-religious atheists, militarists, and racists (tolerance for racists remains the lowest).

---

[63] J. M. Twenge, N. T. Carter, and K. Campbell, "Time Period, Generational, and Age Differences in Tolerance for Controversial Beliefs and Lifestyles in the United States, 1972–2012," *Social Forces* Vol. 94, Issue 1 (Sept. 2015): 379–399.

## Growing in fear, rivalry, and loneliness

*Love is what we are born with. Fear is what we learn. The spiritual journey is the unlearning of fear and prejudices and the acceptance of love back in our hearts.*

—Marianne Williamson

For hundreds of years, oppressors have used fear to maintain power. Fear could be the greatest motivator. Autocratic regimes and religions use it. Many educators and parents who don't know a better way to rein in children's behavior use it. Fear is a favorite tool of abusers, used to keep victims trapped in a relationship.

Buddhists, who have been studying the mind for centuries, say, "The essential cause of our suffering is ignorance." They're referring to ignorance of the true nature of reality. They believe this ignorance makes people crave and cling to an illusory life and that ego thrives on fear.

And we're now experiencing new levels of fear.

When the United Nations issued the *Universal Declaration of Human Rights* in 1948, just a few years after the end of the Second World War, the world was still struggling with the emotional imprint of the recent conflagration and living in fear. In that moment, the United States once again picked up the flag of freedom that inspired the independence of the American colonies. President Roosevelt delivered a speech about the new world order's four freedoms—the freedoms of speech and religion, and the freedom from want and fear. And he made special emphasis of the fourth: freedom from fear. He said freedom from fear required "a worldwide reduction of armaments to such a point and in such a thorough fashion that no nation will be in a position to commit an act of physical aggression against any neighbor—anywhere in the world."

Fast forward seventy years. Many nations, including Afghanistan, Syria, Iraq, Yemen, and Palestine, have not achieved this fourth freedom. And, of course, it's the children who suffer the most.

According to the Pew Research Institute, in 2017 a large percentage of the world's population lived in fear of threats that included ISIS, cyberattacks, the consequences of global warming, and the increasing power of the US, Russia, and China.

In some countries fear comes from rigid hierarchical systems that still fail to recognize the individuality, the emotional needs, or the rights of women and children.

In every part of the world, child-rearing practices and education are often based on fear of authority, fear of ridicule, and fear of failing, which start in grade school and extend through graduate education. Even teachers are in fear—of doing too much or too little, of bad evaluations from their supervisors, of not being in control of the class, of the parents' complaints, and even of children's defiance or attacks.

In my consulting room, I've seen college students experiencing performance anxiety and who could hardly cope with the pressure they felt. I've had adult patients who are still scarred from elementary school experiences and who remember how frightened they were of their elementary school teachers. No, children are not made equal, and not all of them are affected by fear to the same measure. But fear always leaves sequelae; it deeply marks the way we relate to others and the world.

Studies in neurobiology show that early experiences shape the brain of a child and influence its development. Early traumatic events and stress could also adversely affect the organization of brain circuits and the way emotional, social, cognitive, and even physiological responses are set up. Any further strain keeps the primary stress responses activated or makes them easily triggered, affecting our relationships and eventually our parenting. All this gives us even stronger reasons to advocate for increasingly nurturing educational environments.

Because parenting styles establish the affective tone and environment in which a child's brain matures, they will have an enormous impact on the future responses of a person facing social challenges. [64] The way we're raised also sets up the way we regulate emotions and respond to stress.

Observing preschool age children, psychologist Diana Baumrind noticed that children's behavior correlated with specific parenting styles she identified as authoritative, authoritarian, and permissive (1967). Later researchers built up on the list of parenting styles.

Using different sources, I'll briefly summarize the most widely recognized parenting practices.

**Authoritarian style**: sometimes called autocratic, is characterized by a lack of enough emotional warmth and support for the child. The rules are not well-defined for the child (might be inconsistent or unpredictably applied). Norms are strict and reinforced with a because-I-said-so justification. The authority figure is volatile and difficult to please. The level of control is very high. Shouting and physical

[64] A. E. Guyer et al. "Temperament and Parenting Styles in Early Childhood Differentially Influence Neural Response to Peer Evaluation in Adolescence," *Journal of Abnormal Child Psychology* 43(5) (Jul. 2015): 863–874.

punishment are common. Fear is almost a constant and motivates behavior.

**Authoritative style**: usually involves adequate levels of emotional warmth, support, and sensitivity. In this model, parents firmly communicate their expectations, but questioning of the norms is not commonly accepted. The tenet is that parent or teacher "knows better." There might be little freedom of choice and it's still a hierarchical model. Fear of authority is often present, even though communication is usually more open than in an autocratic environment. Punishment and reward are used to motivate behavior. Parents who use this parenting style are more likely to evolve into democratic parenting styles as the child matures and becomes more capable of expressing needs and feelings or of making choices.

**Permissive style**: adults can be warm but may not provide enough or consistent rules or norms (structure). The child might be allowed/pushed to take on tasks or responsibilities before she gets the necessary maturity to undertake them. Neglect might be present. The environment may be chaotic and unpredictable, which can lead to high levels of fear and anxiety in the children. The children don't know what to expect and uncertainty is a constant.

In many cases, parenting styles oscillate between authoritarian and permissive but are still focused on molding child's behavior.

Autocratic and permissive parenting styles are more often associated with family dysfunction. Families might become enmeshed if the boundaries are too permeable or disengaged if the boundaries are too rigid. According to Salvador Minuchin who set the foundations of structural family therapy (*Families and Family Therapy*, Harvard University Press, 1974)), a family becomes dysfunctional when it has trouble adapting to stressors.

**Democratic style**: Rudolf Dreikurs, an American psychiatrist and educator, developed a way to apply psychoanalyst Alfred Adler's theories to parenting and education. Dreikurs said, "A misbehaving child is a discouraged child." Understanding that all behavior has a purpose (to meet an emotional need), he studied ways of eliciting cooperation from the children without having to resort to methods that inspire fear or using a reward-and-punishment approach.

Based on Dreikurs work, Dr. Jane Nelsen developed "positive discipline", a method that "teaches young people to become responsible, respectful and resourceful members of their communities," but also aims at encouraging mutual respect between children and adults (*Positive Discipline*. New York: Random House, 2006). In a democratic parenting style, the rules can be flexible. High levels of emotional warmth and support are the norm. Parents use encouragement and logical consequences (reasonable results that follow behavior). The age of the children and their verbal and cognitive readiness to participate proactively in family life determine how the rules are applied.

Typically, in a democratic family, all family members participate in setting up rules through open discussion of the rationale behind them. Fear tends to be significantly low. The emphasis is placed on caring for the child's emotional need for social acceptance. Achieving self-assertion, competence, inclusion, and courage to face challenges would then be the actual antidote to the so-called misbehavior.

In an autocratic and controlling school system or family environment, the educators or parents tend to feel responsible for providing motivation, disregarding the fact that we were born motivated to learn—that from an evolutionary standpoint, our survival depends on that motivation. Forcing the child to do homework, read certain books, or prepare for tests instead of allowing the child to work at their own pace and pursue their own interests, compromises their interest in learning.

Many parents and educational systems resort to instilling fear (raising their voices, issuing threats, causing embarrassment, punishing, keeping children on edge) as a way to modulate behavior and attain compliance. Dread fosters subordination or resistance and ends up nullifying the child's natural drive to learn. Instilling fear is sometimes disguised as teaching respect, when there is no real regard for the feelings or rights of the others.

Research has proven that punitive measures are not only ineffective in the long term but can make children's behavior worse. They also affect children's mental health.[65]

In a hierarchical society, in which patriarchal structures of power still predominate, discipline is often imposed in an authoritarian manner; children are required to be subordinate to the teacher or parent, and institutions tend to maintain a fear-based structure. Patriarchy, the rule of the father, refers to societies explicitly or subconsciously male-dominated. Males hold significant political leadership, authority and privileges over women, and rule of children.

If adults can't control their anger, if they're unaware of the impact their demeanor and disciplinarian methods have on the brain of the child, they can cause what's known as an environmental trauma. Adverse childhood events (ACEs) that lead to this kind of trauma have been studied

---

[65] D. Smith, D. B. Fisher, and N. E. Frey, *Better Than Carrots or Sticks* (ASCD, Alexandria, VA, 2015).

by the World Health Organization (WHO) and include direct psychological, sexual, or physical abuse but also the child's exposure to substance abuse, mental illness, or criminal behavior in the home.

This matter can't be taken lightly. It has long-term consequences on the child's ability to regulate his or her emotions. Violent displays of anger make children feel threatened. Many researchers agree that what we call difficult behaviors are very likely responses to stress, reactions of a child who is feeling unsafe. Instead of expressing their fear, children more often express anger, an emotion that conveys some sense of power.

> *This single mom couldn't understand why her son was not responding to her constant reprimands, restrictions, and punishments. He didn't seem to listen. He wouldn't change his attitude or behave any better. They were constantly fighting. So, they came for therapy. You could almost see and touch the wall growing between them. He was starting to display risky behaviors, the mom explained, and she was right to be worried. But she was making a mistake many parents and teachers make, a very common mistake. She was expecting positive results from her disciplinary methods while disregarding how ineffective they were. When the son's behavior didn't change, she just amplified the exact same disciplinary method that was not working. And no matter how much she amplified it, it still didn't work. It was like insisting that someone take more of the same medicine that has proven to be ineffective or toxic. We discussed how autocratic forms of discipline lead to hostility, which perpetuates a cycle of domination, fear, humiliation. It took a while for her to admit her son was no longer a little child and new rules and norms, appropriate for his age, were needed. I saw her again three years later and her report was more than positive. By changing her perspective about parenting and their relationship, by loosening her grip to let him grow, her son felt valued and trusted, and then became more willing to modify his behavior.*

Do you remember how it felt to be a little one? We need to be able to see the world from the perspective of the child and to provide empathic and nurturing responses. If the adult (parent of teacher) responds appropriately to the child's emotional states (fear, anger, frustration), he or she will help the child learn how to self-regulate and communicate emotions.

If the definition of success we teach children is based on achieving recognition, notoriety, and riches, on gaining power and status, they'll likely develop both competitive behaviors and fear of failure. Rivalries will become evident both inside the family and at school. In a very intuitive way, little children see peers and siblings as friends, but as they grow up, they learn to see peers as competitors, as rivals. In the long run, power struggles will involve the teacher or parent, and the child might become confrontational to figures of authority. It's not uncommon to observe a teacher exasperated and engaged in an unhealthy and fruitless argument with a child.

With no intention to stereotype, I'd say that Latin American mothers tend to be overprotective (too present) and fathers/providers tend to be peripheral (too absent). In a systemic family model, we see these families as enmeshed or undifferentiated. Parents expect children to grow to be like them. There is little independence or privacy. If the mom believes authority must come from a male, she will refrain from discussing or applying rules, limits, and boundaries. If the father is absent, the mother might let the older children shoulder some parenting responsibilities. This generates even more dysfunction, conflict, and apprehension.

Mixed families and those in which the heads of household are divorced or single are now more common than ever, each new composition of a family contributing their own share of struggles, risks, insecurities, and fears.

When homes break up, schools might not be able to provide much help, even if they have a counselor. Children might come to school with unconscious expectations that their affective needs could be fulfilled by teachers or other adults. The child might project on the teacher his unsatisfactory parent-child relationship. (This could manifest as dependency and submission or rebellion against authority.) Many teachers care deeply about the well-being of the students. But if teachers are focused on the academics and the behavior and are unaware of the underlying psychological dynamic taking place in the student's personal life, they can't offer answers or design strategies that could contribute to their emotional growth.

The training of teachers seems to be focused on literacy, math, and academic discipline. Some training includes topics in cognitive development and learning styles. However, teachers are seldom trained in how to respond to the child with empathy or provide comfort. Sometimes they're just not equipped or are too overwhelmed to offer a more nurturing environment.

Unfortunately, many school systems tend to maintain and reproduce intimidating structures. Teachers might be inclined to mistrust the student. (Is he cheating? Lying? Mocking me? Would he be able to do his work if I weren't hovering over him?) They might also fall short in creating intellectually stimulating environments, promoting cooperation, or offering emotionally safe environments. They might assume that their role is limited to passing on information and helping the kids succeed academically.

I have known many teachers who are truly compassionate and honestly care. Unfortunately, very often the system forces teachers to use examinations at all times (quizzes, oral assessments, standardized written tests, written papers) that generate significant stress. In most cases, the goal is to help students memorize information, not process it, while homework fuels performance-based anxiety and causes stress, not only to the children but to the whole family. We need to remember that what separates memorization from learning is the acquisition of knowledge and meaning. A nurturing environment, not fear, facilitates the learning process.

Because emotional issues have basically no room in a classroom (even less in high school than in grade school), instead of developing the ability to give and receive affection, children learn to operate according to the every-man-for-himself philosophy.

But as Daniel Quinn said in *My Ishmael* (Bantam, 1997), "Tribes survive by sticking together at all costs, and when it's every man for himself, the tribe ceases to be a tribe."

An antagonistic environment generates fear and compromises the feeling of safety. If hostility and mutual distrust between young people and adults prevail, the motivation to learn disappears. Too often we try to modulate behavior by using rewards and punishment. But also, and without being very aware of the consequences, children are encouraged to engage in competition, which promotes rivalries between siblings and between children and their peers.

Parents and educators need to focus on stimulating the inquisitive minds of the children and supporting the acquisition of skills that facilitate learning, expression, and creativity. However, the emphasis is more often on the outcome than the process; grades, achievements, and awards are used to measure children's "success."

The goal becomes winning over others, being on top, becoming the best of the group. Without intending to, we prompt performance anxiety and a fear of failure. Without knowing it, every time we push children to seek an award, we are contributing to the reinforcement and constant activation of the dopamine-producing system (of which we'll more talk later).

And we do little to promote empathy and compassion.

Our system of rewards also flatly fails for a reason we sometimes won't acknowledge: we incentivize the child with what we had previously restricted or prohibited. For example, if he shows us an achievement, we allow him access to the video game to which he's becoming addicted, or we buy him junk food, which we usually don't want him to eat. By rewarding him with what is, by norm, forbidden, we're stimulating cravings. The banned object will now be associated with pleasure. We also fail when comforting a distressed child with, for example, a lollipop. Instead of displaying empathy, we're mindlessly teaching them that food (usually sweets) helps us manage our feelings.

Another aspect to consider is that while it's healthy to encourage autonomy and responsibility from an early age, we must be careful not to encourage individualism and egotism. This search for individual freedom, which seems to be a sign of our times, derives from the struggle for civil liberties that the Greeks advocated for three thousand years ago. Now it's become an important pursuit for the modern individual in Western countries. It's one thing to breed autonomous, responsible and self-disciplined individuals' it's quite another to foster self-centered individualism, which becomes the source of feelings of loneliness and isolation. It doesn't stimulate interconnectedness. We'd be better off devoting efforts to fostering interdependence and learning mutual support.

Children, and especially adolescents, constantly receive conflicting messages about breaking away from the pack (autonomy) and socializing with others. Loneliness results when attachment needs are not fulfilled.

In 2011, the United Nations Children's Fund (UNICEF) released a report concluding that among industrialized countries, England was the worst a child could grow up in. It had very low indices of well-being among children and high inequality. UNICEF then commissioned IPSOS Mori's Social Research Institute to dig into their findings a bit further. Among the conclusions of the new study was that "British parents have a compulsion to buy things for their children, but do not spend quality time with them." The consequences? Great Britain has the most stressed, unhappy, and sedentary children in the developed world.

Experts point out that children, who are the most vulnerable to commercials, spend too much time sitting in front of televisions, game consoles, and computers without supervision. To make things worse, game and television content foster rivalry, violence, and fear. Oh, but yes, the children "connect" through war video games!

Children are growing up in environments with fewer rules, less accountability, and little dialogue. They're engaging more often in self-destructive (dysregulated) behaviors—cutting, addiction to drugs and gaming, early and promiscuous or unprotected sex, gluttony, debt-based consumerism.

What are the social implications of parenting styles that promote competition instead of empathy or cooperation? Parenting styles that opt for repressive methods and fear in an attempt to modify children's behaviors while at the same time stimulating the search for self-sufficiency?

## Fear drives out love

Here's a story about choice and potential. What are we born with? What tendencies do we feed?

> *A popular story of unknown origin says that an old Cherokee is teaching his grandson about life.*
>
> *"A fight is going on inside me," he said to the boy. "It is a terrible fight and it is between two wolves. One is evil—he is anger, envy, sorrow, regret, greed, arrogance, self-pity, guilt, resentment, inferiority, lies, false pride, superiority, and ego." He continued, "The other is good—he is joy, peace, love, hope, serenity, humility, kindness, benevolence, empathy, generosity, truth, compassion, and faith. The same fight is going on inside you, and inside every other person, too."*
>
> *The grandson thought about it for a minute and then asked his grandfather, "Which wolf will win?"*
>
> *The old Cherokee simply replied, "The one you feed."*

Neuroscientists believe that because of the design of our cerebrum, we're doomed to experience a negativity bias. The brain seems quicker to perceive danger than opportunity. Recalling positive occasions and staying grateful, confident, and optimistic requires a conscious effort. Traumatic stress can permanently affect various brain areas, including the amygdala, hippocampus, and prefrontal cortex, in such a way that after experiencing a trauma, the brain circuits that respond to threats can be easily triggered either by memories or by sensory inputs of anything that resembles the original trauma.

The good news is we could actually train our responses to perceived threats and choose which of three different systems we allow to be activated by stress. More about this later.

The fear propagated by sensationalism, Hollywood movies, the swings of the stock market, or terrorist attacks will reinforce our negativity bias, enhancing our dread of an uncertain future. This is based on genetic wisdom, designed to protect us from harm.

You might wonder if movies contribute to spreading, reflecting, or explaining fear. French author and political scientist Dominique Moïsi has said that in time of globalization, television series have become a cultural reference (if not the only one) essential to the analysis of "the emotions of the world." He believes TV series have become the equivalent to the potboilers of the nineteenth century, which were an essential component of the culture. In an interview with Fernando Afanador (Revista Semana, February 2018) Moïsi said, "And scriptwriters—their best—are comparable to the great novelists of the past like Balzac, Flaubert and Dickens. They do not settle to coldly analyze reality: they perceive it and guess it, thanks to the power of their intuition and the courage and lucidity of their imagination. In fact, these writers have become the best analysts of society and the world today, not to mention the most reliable futurists." (Quote translated by the author.)

Our desire to possess more and more *stuff* is also fueled by fear: a primeval dread of not having enough should a tragic event strike us. The more uncertain we perceive the future to be (and the more we suffered from scarcity in the past), the more likely hoarding will be associated with feeling safe. Publicists know this.

Fear motivates consumerism. "Fear appeal" is the name for a set of marketing strategies designed to create uncertainty and doubt, inducing us to take certain actions. Although it's an immoral practice, corporations usually employ fear appeal in marketing. Yes, of course, if you know your customers are vulnerable to a threat, you need to warn them.

Fear appeal strategies are also used to motivate you to avoid risky behaviors (drugs or unprotected sex, for example). I'm talking about manipulating our negativity bias to make us buy products and services presented as safety essentials, even when the actual danger they prevent is negligible.

Pay attention the next time you watch a commercial. Does it warn of the possibility of being socially ostracized if you don't smell good or if you have halitosis, for example? Does it emphasize safety features when selling us a car or a car insurance?

I often say that our three worst enemies are refined sugar, fear, and addictions. The three of them are antagonists to love.

Refined sugar seems to affect production of serotonin (that modulates mood and social behavior, appetite and digestion, sleep, memory, and sexual desire). It also creates a physiological effect similar to that of stress (it even raises cortisol levels). Addictions, including to alcohol, interfere with love, for various obvious reasons. And by activating the primary fight-or-flight response, fear blocks the expression of other emotions, such as empathy, and slows down the activity of the cerebral cortex associated with logical thinking. This would explain why behaviors motivated by fear are often irrational.

In the process of creating automatic responses to fear, the brain learns to efficiently assess signals of a threat captured by our senses. Involved in this process are the amygdala, the sensory areas in the cerebral cortex (where stimuli are processed); the brain stem, which controls the reflexes and autonomic responses; and the pituitary gland, which secretes factors into the blood that act on the endocrine glands to either increase or decrease hormone production.

Since it activates a stress response that makes your body produce adrenaline and cortisol, fear becomes harmful to the physical body, and especially if it's sustained. As mentioned before, a stressed body increases the presence of free radicals, which accelerates the aging process.

Fear can literally paralyze and kill, as happens to the rabbit that feels threatened and collapses. A person suffering from a panic attack can have symptoms very similar to those of a heart attack: the heart speeds up, breathing becomes irregular, and we may even lose consciousness.

But the worst effect of fear is that it obstructs the possibility of love. Fear make us distrustful and defensive. It fuels our jealousy and motivates vengeance. Fear is not only the opposite to love, it's also a weapon that has been used by abusers and oppressors. And fear and its partner, anxiety, are what most often bring patients to psychotherapy.

## Consumerism, fashion, and the need to belong

Karl Marx wrote about an active relationship between man and his environment. As early as 1844, he wrote "Man *lives* on nature – means that nature is his body, with which he must remain in continuous interchange if he is not to die. That man's physical and spiritual life is linked to nature means simply that nature is linked to itself, for man is a part of nature" ("Estranged Labor," *Economic and Philosophical Manuscripts of 1844*).

The consciousness of a person is determined by their participation in what is called a "productive life," which refers to the social conditions in which we live and the spiritual representations we derive from those conditions. Our viewpoints, our lifestyle, the decisions we make, are products of a society presently ruled by consumerism. Think about how many people cover their basic needs with a fraction of their earnings, but the rest of their income goes to feeding the pleasure of owning.

This is an intriguing behavior. We hire nannies so we can leave the house to go to work so we can pay for a nanny. We spend little time with the children because we work so hard to be able to provide for their needs, but more than anything else, to deliver what they want. We accumulate objects over the years, too afraid or lazy to discard them, even when they no longer have any use. Writing for curbed.com, reporter Patrick Sisson wondered if we're becoming a nation of hoarders. In 2014 there were 48,500 self-storage locations across the US, generating a yearly revenue of twenty-four billion dollars.

In this context, the media has become an ideal tool with which to control the minds of consumers in an economy that, in addition to producing essential items, pushes us to work twice as hard to buy non-essential items. Our range of needs transcends the real ones and expands to include those that are actually desires created by our psychological situations.

Consumerism targets the unsuspecting person who comes to believe that the accumulation of certain goods and services is necessary for their proper social insertion. Consequences don't seem to matter. People are happy to embrace the superb invention of a little plastic card that makes indebtedness completely painless—until you realize the magnitude of the obligations in which you've been trapped. It's even worse now, when, with a simple tap on your smart phone, you can buy something before you have time to ask yourself if you really need it.

The consumerism society is as old as the industrial revolution, but today, hyper-consumerism drives the economy, and what we buy defines who we are.

In her excellent documentary *The Story of Stuff,* Annie Leonard quotes Victor Lebow, an independent market consultant, economist, financial consultant, and author. Shortly after the Second World War, Lebow analyzed the needs of the market and suggested the following winning strategy to corporations:

> Our enormously productive economy demands that we make consumption our way of life, that we convert the buying and use of goods into rituals, that we seek our spiritual satisfactions, our ego satisfactions, in consumption. The measure of social status, of social acceptance, of prestige, is now to be found in our consumptive patterns. The very meaning and significance of our lives today expressed in consumptive terms. The greater the pressures upon the

> individual to conform to safe and accepted social standards, the more does he tend to express his aspirations and his individuality in terms of what he wears, drives, eats—his home, his car, his pattern of food serving, his hobbies.
>
> These commodities and services must be offered to the consumer with a special urgency. We require not only "forced draft" consumption, but "expensive" consumption as well. We need things consumed, burned up, worn out, replaced, and discarded at an ever-increasing pace. We need to have people eat, drink, dress, ride, live, with ever more complicated and, therefore, constantly more expensive consumption. The home power tools and the whole "do-it-yourself" movement are excellent examples of "expensive" consumption.

In 1955 Lebow already seemed to understand well what a powerful weapon television was for dominant producers: "Television achieves three results to an extent no other advertising medium has ever approached. First, it creates a captive audience. Second, it submits that audience to the most intensive indoctrination. Third, it operates on the entire family."[66]

Branding is one of the most critical business strategies used to position a product in the market and promote consumerism. According to *Forbes* contributor Jerry McLaughlin (December 14, 2011), "Brand is the perception someone holds in their head about you, a product, a service, an organization, a cause, or an idea. Brand building is the deliberate and skillful application of effort to create a desired perception in someone else's mind."

Psychologists, sociologists, and economists analyze in which way and why the opinion of others modify our behaviors. Their conclusions help them find the best way to successfully use certain colors, images, and sounds to speak to our emotions and appeal to our minds. That is what advertising and marketing companies are for. The success of branding demonstrates how advertising manages to influence the perceptions, behaviors, and consumerist habits not only of the housewife but especially of young people and children. Fashion reinvents itself periodically in synchrony with advertising.

Advertising is a specialized profession, hundreds of years old, that competes with education and even with religion as a shaping force for desires, behaviors, and identities. According to the business data platform Statista, in the media alone, more than five hundred billion

[66] Victor Lebow's full text can be found at: notbuyinganything.blogspot.com.

dollars were spent in 2018 on advertising worldwide, and a fifth of this money went to Facebook and Google.

Gossip, bullying, criticism, and isolation also become modulators of social attitudes, defining what is socially acceptable and what is not. In this way, fashions and uses that contribute to determining what's suitable for a social group get strengthened. I dress and decorate my space according to who I think I am, what I want to be, and the image I want others to perceive. Our need for belonging and inclusion puts us in the dilemma of resembling others to get accepted by the pack—or to look unique, which requires particular strength, since we risk isolation from people we like. However, the bonds created through fashion imitation (behaviors, music, clothing) are too superficial to last. In the world of today, the immediacy of the object or activity that satisfies you has a higher value than love, which might get sacrificed in the process. We also compromise natural resources (we show no compassion for the planet or those affected by our waste), and we upset the body. Fashion is costly.

The problem with consumerism is that it makes an object of the human being. Even at a personal level, love is deformed, transformed, misunderstood, and misinterpreted. Women need to be pretty to be loved, for example. Women resent becoming objects of desire and struggle to become the subjects of desire instead, but they might also end up objectifying others. In this context, sexuality doesn't contribute to connecting the lovers. Men resent knowing they might be attractive primarily because they're potentially good providers.

A shoe is no longer produced because it's good to walk in but because it's a decoration. That's how it generates profits. Advertising is deceptive: products are presented to us with the argument that we *need* to own them. Commercials seduce instead of informing about the product. It's not only objects that are advertised. In the economic system prevalent in the world, everything is merchandise, including kindness and beauty, and even health and knowledge.

The effects of consumerism on people's minds and traditions are devastating. In a world ruled by fear and a constant search for gratification, it seems that the only thing that could save humanity would be an increased sense of duty, an inclination to service, and a notion that the common good comes first.

In 1974, German professor of communications and pollster Elisabeth Noelle Neumann developed the spiral of silence theory, about the growth and spread of public opinion. It's based on the assumption that fear and isolation make us think, feel, and express ourselves as the majority do. She believed we had a kind of sixth sense that allows us to scan the environment and follow what's trending, and she blamed the media for shaping the public mood. Even though her ideas have been controverted in a world that honors individualism, we know, better than ever, that those who want to shape public opinion strive to control the media. Our need to belong explains the success of social networks as communication outlets defining what's trending now. Belonging to a group provides us with an identity and offers security, which is an evolutionary advantage. However, while a group tends to be supportive of peers, it recognizes strangers as potential adversaries.

## What a frantic life!

*The danger of the past was that men became slaves. The danger of the future is that men may become robots.*
—Erich Fromm

In almost fifty years in practice first as a medical doctor, and then as a psychotherapist, Reiki teacher, and Trager and Qi Gong practitioner, I have observed the urgency with which people expect healing and symptom relief to occur. Most people look for an immediate pain-killer effect. Grown accustomed to quick results from popping a pill, many people expect to heal with minimal effort on their part. Some even go for invasive, traumatic procedures instead of opting for lifestyle changes that, even though they require some discipline, would bring about long-lasting benefits.

There seems to be an urge to get fast results, to save time, to absorb life (and with a minimum effort). Many technological advances respond to that urge: elevators and moving sidewalks are good examples. Bicycles, motorcycles, and automobiles run faster than ever. Hypersonic airplanes are now capable of flying at five times the speed of sound (about thirty-eight hundred miles per hour). We also invented fast food, fast fashion, fast checkouts, express delivery, instant coffee, and speed-dating. Since we now have access to digital movies, it's become easier to indulge in binge-watching. And we communicate instantly with anyone around the world.

But the same technology was supposed to make our lives easier, to free time for leisure. Merriam-Webster defines leisure as "freedom provided by the *cessation of activities*" (italics mine). In Buddhist terms, leisure might mean time for meditation and contemplation. Instead, we pack our free time with more and more activities.

We're sometimes unaware of how the fast pace at which we're living increases our stress levels and prevents the body from properly repairing, regenerating, and healing tissues. Therefore, our bodies tend to be in a state of constant inflammation.

Another side effect of living at this crazy pace seems to be low-frustration tolerance and a need for immediate gratification. Our youngsters talk and think so fast that they don't have patience for slow oldies like me.

I remember this woman who injured her knee playing squash. Even though she kept hurting, she wouldn't stop playing. She avoided treatments and rehabilitation therapies. "When and if the knee stops working, then I'll go for a knee replacement," she said. She was only thirty-five.

It's sad not only that we've come to think of the human body as a set of parts that can be repaired or replaced but that we won't slow down or stop hurting ourselves for health's sake.

Each culture has its sense of time. The axiom "time is money" is especially true for the United States and Japan. These are societies oriented toward profit and productivity; leisure is inconceivable, at least during the productive ages. Cultures that are future orientated have different time management styles than those that are present oriented, like India and Spain, or past oriented, like England.

For example, a fisherman in a village in Northern Colombia would have a different sense of time than an American executive. If he leaves with his *atarraya* (cast net) at five o'clock in the morning and returns around ten, his labor finished but with only a few fish, did he waste his time? If he spends the rest of the day with his family or helping out with domestic chores, is that lost productive time? And if for a Russian an accomplishment is to get a good remuneration for a job in which he didn't have to spend but a few hours of work, is the rest of his time, which he dedicates to socializing with others, a waste? Are the Spanish, French, and Italians, who still close their businesses at lunch time, just losing sales?

Interestingly, for maybe a hundred years, people in the United States have explored how to compensate for their frenetic lifestyles with sedative practices imported from India and China. However, many of the philosophical and contemplative aspects of Eastern practices, like yoga, Qi Gong, or meditation, are left aside. Yoga, for example, has evolved into exercise for physical fitness. Teachers and students often neglect the soothing benefits of a practice that can train your mind to focus, can reduce stress, increase awareness of the body, and attune the individual with the universal energy. Yoga and Qi Gong can actually stimulate brain circuits involved in the regulation of emotions, giving us a more placid attitude toward life: a benefit that would extend to others. Neuroscience tells us that, if you are calm, you bring me calmness.

We often say, "I have no time for anything." But what we actually mean is, "I don't allow myself to take pause." Our calendars are full. While the goal to experience life to its fullest is valid, the paradox is that we keep packed schedules that go hand in hand with a sedentary life. We engage in work and activities that become addictive.

In "Economic Possibilities for our Grandchildren" published in *Essays in Persuasion* (New York: Harcourt Brace, 1932), the famous economist John Maynard Keynes predicted that capitalism would progress so much that his grandchildren would not work more than three hours a day and surely for pleasure, not out of necessity. What was Mr. Keynes thinking? So far, erroneous prediction. In the corporate world of today, people work without schedules around the world and not exactly for pleasure. And the devices that technology has created to help us save time very often do the opposite. Maybe these tech advances allow us to do many more things, but the result is we're busier than ever.

The digital world has imposed quite tedious tasks on us, from erasing junk messages from our email accounts to answering what turn out to be robocalls or replying to endless and sometimes futile text messages arriving at any time of the day or night. And what about the time spent trying to understand how to navigate the ever-changing online platforms we use to pay bills or do other tasks, or waiting on the phone for customer service when our computer or internet service needs their support?

The speed of our digital devices has seemingly made us anxious, very aware of time, feeling the clock pacing our steps. The result? High levels of stress and lost equanimity—that freedom from reactivity that Buddhists consistently seek.

The principal of my grade school often accused our fathers (who wouldn't attend PT meetings) of "living to work instead of working to make a living." Maybe more parents attend PT meetings these days, no doubt, but now keep hooked to their phones.

People with a clear vision of their responsibility regarding the future seem like they might grow wings to fly to new destinations and face tough challenges, but they often sacrifice the quality time they could be sharing with family or friends. People who are devoted to caring for the planet or serving the community might be sleeping fewer hours, eating quickly and mindlessly, and not reserving time to decompress. This frenzy compromises the expression of self-love, love for others, and care for the planet. Maybe we could learn the Buddhist's middle way of moderation.

Examining in depth all the obstacles we find in the way of becoming empathic, compassionate, loving human beings needs to be the subject of a whole new book. For now, I scrutinized the obstacles I feel require more attention. My hope is that, once persuaded that we are physiologically equipped for love, as we will discuss in the next section, we'll be motivated to take both personal and social action to overcome those obstacles, anticipating the personal, community, and universal advantages of a more supportive world.

# PART THREE: A Science of Love

*If our brains were simple enough for us to understand them, we'd be so simple that we couldn't.*
—Ian Stewart

## Why should we learn neurobiology?

With my past books, professional work, and classes, I've tried to make a case for the importance of learning about the body and how this learning could become a key to good health.

I've tried to understand health and disease from an integrative perspective since the time I was a medical student. After graduation, I thought I would work as a general practitioner or a family or community doctor rather than specializing, fearing that, as I think happened to some colleagues, I could end up focusing on a single bodily system or part, while losing the perspective of the individual as a whole.

In Colombia you don't get the final signatures on your MD diploma until you complete at least a year-long mandatory rural practice. I fulfilled this requisite by working as the director of a medical center with a few hospital beds in Barrancas, in 1972. Located in La Guajira Peninsula, in the northernmost part of Colombia, the town is now well known for the nearby El Cerrejón, a large open-pit coal mine that represents one of the biggest earthmoving operations in the world. The town grew up to become quite a modern urban center. But when I practiced there, the coalfield was inactive, and Barrancas was a small municipality with about three thousand souls, surrounded by *rancherías* inhabited by people from the Wayúu tribe. Those rancherías are hamlets made up of small dwellings with bahareque walls and ceilings made of *cardón*, a resistant wood that keeps the house warm day and night; one or two pens for goats and eventually a cow; a well; a *jagüey* (a natural deposit of rainwater); and a family cemetery. Often, patients came to their appointments with relatives in tow. If they felt grateful, they would invite me to visit their homes, eat their food, and attend their celebrations.

I was a twenty-two-year-old city girl with great ideas and little experience in life. Soon I realized how important it was for me to learn more about Barranca's culture and the environment in which patients lived if I wanted to have a broader perspective of their maladies and get them to comply with treatments. I began to offer nutritional advice (based on their dietary preferences) and strategies to offset stressors contributing to elicit symptoms. And I found ways to engage the patients' support systems in their treatment.

However, it was only many years later—after I quit my medical practice to specialize in art psychotherapy, after I learned Reiki and later studied massage, Qi Gong, and somatic therapies, after reading many books and learning many lessons—that I finally arrived at a fair understanding of the multidimensionality of the body and of how everything in it is connected. Now I know, for example, that all organs (not just endocrine glands or the nervous cells) communicate with each other.

Candace Pert's book *Molecules of Emotion* (Simon & Schuster, 1999) helped me understand the information network within the body. It has two major components: the messengers (neuropeptides, neurotransmitters, hormones) and the cell membrane receptors that pass the message on to the cell. I learned that emotions talk to us through the body and the body talks to us through activating specific regions of the brain. Or as Pert put it, "The body is the mind."

Eventually, I gained an expanded understanding of how the environment, genetics, emotions, and lifestyle influence our health. I started to see physical and mental symptoms as body alarms and understood the body's immense capacity to heal, recover, regenerate, and repair tissues, given favorable conditions. I have shared this understanding of the multidimensional human body in my book *Regaining Body Wisdom* (Eyes Wide Open, 2008).

We hastily delegate the care of our bodies to our physicians when we don't feel well. We might think that this is what needs to be done, since the doctor is the *expert* and as such, they should be in charge of the patients' health and prescribe what to take or do. Unfortunately, doctors seldom involve patients in their own care. Medicine still needs to evolve toward a more patient-centered paradigm that addresses the patients' concerns, lifestyles, and beliefs and expands their understanding of the symptoms, allowing them to take responsibility for their health. This is the main rationale for including this section in the book. But I warn the reader that it's almost impossible to make it an easy read. Summarizing the sheer volume of available information about how the brain works in a few pages is an ambitious undertaking, but at least you'll get acquainted with some key terms and basic concepts.

In the West, a noteworthy popularization of complementary and alternative medicine (CAM) practices in the past few decades has led to a greater openness to new paradigms among those in the medical establishment. However, if healthcare practitioners look for better treatment outcomes and adherence to treatments by patients, they must become better patient educators, and use a more holistic approach to help improve the patients' understanding of their conditions. For example, psychoeducation allows us therapists to better support and empower our clients.

Doctors are trained to prescribe pharmaceuticals or surgery as first options, mostly because in many cases drugs provide you with a quick, though temporary, fix. Medication can seem a more efficient way to treat a symptom or condition than attempting to engage a patient in healthier ways of living.

I often tell my students this story of a moment that changed my life. I'd been a heavy smoker since I was sixteen, but I quit craving cigarettes in 1993 all of a sudden after receiving a healing session with a medical doctor who had embraced Taoism. I didn't believe in "energy healing," and went to him just out of curiosity. As any person addicted to nicotine, I'd had difficulty trying to quit smoking, so I was struck by this sudden and permanent transformation into a nonsmoker. It led me to eagerly explore new paradigms of healing and to embrace a radical lifestyle change. Food, exercise, meditation, and Reiki have been my only medicines since then. When my body gives me a warning, I double-check my nutrition, habits, and possible unprocessed emotions. Some foods compensate for nutritional deficiencies caused by stress or faulty diets, others fight inflammation, relieve pain, or prevent acid reflux. I take supplements only occasionally—maybe a homeopathic remedy for a cold and some vitamins when it feels necessary. I try not to skip my daily walk, and practice Mentastics, a self-care movement component of the Trager Approach. It has helped me be more aware of my body and my emotions and stay in good shape, while preserving flexibility and balance, which are essential if you are concerned about falling, breaking a bone and compromising your independence in old age. I've found that staying healthy requires an expanded understanding and awareness of the body. I've learned that a healthy nutrition, movement routines, stress management, connecting with others and the earth, and mindful emotional regulation, all impact my health.

In this modern world, where we pop a pill before we change a habit, not listening to the body or learning about our physiology stops us from taking timely action to prevent disease. To recover and maintain our body's wisdom, we must not only be mindful and listen to our symptoms but understand how our body functions and learn what our normal physiological responses are.

Freedom comes from knowing. Someone said wisdom is the combination of knowledge, experiences, and insights. So, let's get some knowledge first: learning about the anatomical and physiological substrate of emotions such as fear, anxiety, rivalry, love, and compassion is a first step toward improving our mental health, our capacity for love, and the way we interconnect.

As said before, neuroscience is, of course, a far more complex topic than I could ever elaborate in a few pages. But I'll attempt to introduce a simplified version of the functioning of some of the aspects of the nervous system's circuitries associated with emotional regulation, stress management, and empathy. My purpose is to establish the idea that we're born physiologically equipped for empathy and love, and we just need to stimulate its development.

Our very human problem is that when fear and reward-seeking behaviors prevail, neuronal circuits associated with aggressive, evasive, competitive, or addictive behaviors will stay activated. But we can choose to turn on the circuits associated with kindness and solidarity and to do it at will.

Now, before we start talking about the brain's circuits, let's briefly explain how scientists study the brain. Before the advent of neuroimaging, the main research methods were clinical (observation), electroencephalograms (EEG), behavioral tests, and microscopy.

In the last three decades, neuroimaging, which includes scanning the brain with magnetic resonance (fMRI) and positron emissions (PET), has proven an invaluable tool for neuroscientists to use in studying the mechanisms of love. For example, it has helped us discern how different types of love—passionate love, brotherly love, maternal love, or unconditional love—activate different regions of the brain. Event related potentials (ERP), which uses electroencephalography and magnetoencephalography to study infants, children, and adults' electrophysiological responses to a stimulus, has allowed researchers to learn more about developmental processes. The quantification of hormones and neurotransmitters (neurobiological markers) released when a person experiences a particular emotion is still in its initial stages, but even with only limited measuring devices, scientist have been able to identify chemicals that are released when we experience certain emotions or states of mind such as anxiety, depression, focused attention, obsession, and possessive jealousy.

One limitation of neuroimaging studies is that they only reveal areas that activate in the brain when an individual is completing a task, but not necessarily which areas are critical to performing the task. Therefore, many neuroscientists still complement their studies with the old method of correlating brain lesions with the cognitive, emotional, sensory, and motor functions affected, to map the brain and identify which regions are necessary for a specific task or associated with specific behaviors and emotions.

Many people don't yet understand how the past determines many of their current behaviors, reactions, or feelings. They don't know what imprint they've left in their brains. I've found it useful to educate my patients in basic neuroscience concepts; it makes their therapy more effective. I facilitate their learning about the psychobiology of trauma, the emotional and visceral brain, the vagus nerve (involved with how traumatic experiences affect us), what makes the memory of a trauma different from other memories, and what factors protect us and allow us to recover after a traumatic event. This allows clients to be in better control of their own mental processes. We even discuss how diagnoses listed in psychiatric manuals (none of them linked to physical tests or biological markers except for dementias) can easily become a negative label that affixes to our identity, preventing us from achieving change.

I've also found it useful to discuss the concept of neuroplasticity in counseling sessions, because it brings hope to know that the brain is alive and adapts and changes, no matter our age; therefore, it's not our destiny to chronically suffer from the same symptoms. And it's encouraging to know that, no, we're not bound to repeat the history of our relational patterns. Science can now prove it.

My mom used to repeat an expression equivalent to the English "you can't teach an old dog new tricks." It prevented her, for example, from daring to take lessons in drawing, for which she had a talent. She felt it was too late for her. According to mainstream "knowledge," the adult brain couldn't change or regenerate, and new neurons (nervous system cells) wouldn't grow in the brain after a certain age. We were wrong. In 1992 professors Perry Bartlett and Linda Richards discovered stem cells in the mouse brain, which suggested that adult individuals could actually grow new neurons. We know now that neurogenesis (the formation of new nervous tissue) occurs in at least three regions of the mammalian brain: the hippocampus, the subventricular zone, and the amygdala. We also know now that neurons create new neural pathways to communicate with each other and that these pathways can regenerate and grow more connections (synapsis) through life. New learning, as well as our behaviors, shape our brain, no matter our age.

When Ginette Paris talks about people's painful experiences in her book *Heartbreak: New Approaches to Healing—Recovering from Lost*

*Love and Mourning* (Mill City Press, 2011), she says that although memories of a traumatic experience can't be crossed out or erased from our brains, we can indeed overwrite them by adding new experiences and responses that will change our perspectives, emotions, and behaviors. Paris says, "The memory of a traumatic experience (like the loss of a partner) cannot be erased from the folds of our brain. In other words, trying to forget the relationship, its beauty, or the pain of having lost it, doesn't work. At the brain level, it's only possible to add new skills to the repertoire of answers, and these new answers will eventually take the place of the previous ones."

Understanding the above seems to help my patients. If they have attached a previous diagnostic label to their identity, they are more susceptible to worry and to feeling the stigma of mental illness, and information provided by a therapist might take more time to sink in. However, discussing neuroplasticity, understanding that the brain can change, "feels like a sentence has been commuted," one patient told me. It's reassuring to understand that symptoms are adaptive responses (a reaction to challenges presented by the environment at some point in time). It's hopeful to know that biology doesn't define us or determine what we can be or do. We have choices when it comes to responding to an event, and our reactions and responses could be modulated and changed with committed work.

I remember experiencing some frustration in the past while exploring possible changes in a client's parenting strategies. The client said, "There's not much to do. The doctor told me my child suffers from such-and-such disorder." Psychiatric diagnoses can leave a mark that is difficult to erase; they can even make a parent resign their responsibility to improve their relationship with the child.

It's important to note that the way a human being responds (physiologically and behaviorally) to the challenges presented by the environment depends on factors such as early experiences, social environment, and even the climate.

The discovery of the correlation between our experiences and our neuronal architecture has contributed to changes in psychotherapy approaches. Our increasing understanding of how our outlook on life, our behavior, our mood, and the way we regulate emotions are the product of a combination of genetic factors, our previous relationships, our work, our learning process, the decisions we make, and the environment certainly creates room for hope. Discerning that the brain is plastic, moldable, and doesn't stop changing throughout our lives leads us to focus on learning how to make conscious choices that will modify not only our behavior or the way we feel but the course of our lives.

In *Parenting from the Inside Out: How a Deeper Self-Understanding Can Help You Raise Children Who Thrive* (Penguin Putnam, 2005), neurobiologist Daniel Siegel and coauthor Mary Hartzell do a good job of explaining how relational patterns—learned or acquired through parenting, school experiences, family relationships, reading, watching movies, hearing songs, etc.—define the type of bond we tend to establish with partners, children, families, communities, and societies. And these patterns, in turn, determine the structure and function of our brains.

From an evolutionary point of view, our ability to cooperate has marked the search for increasingly sophisticated forms of communication and, at the same time, has contributed to the development of the current human brain.

It takes time and persuasion for a person to accept that diagnoses are not final, that even if environmental trauma occurred, a change in the way a family interacts, supported by a therapeutic relationship, can help them heal. When parents understand how the brain works (the science of parenting), they can better soothe their children, model empathic behavior, become aware that they might be experiencing "blocked care," and then they can find ways to engage in a healthier relationship with their children. In *Brain-Based Parenting: The Neuroscience of Caregiving for Healthy Attachment* (W. W. Norton, 2012), Daniel Hughes and Jonathan Baylin brilliantly explain the above concepts.

Mine is not a behavior-centered approach to therapy, but mostly a client-centered pragmatic and holistic one. I support clients in their process of gaining awareness of their choices and responsibilities, repetitive and limiting relational patterns, and the obstacles that might make them feel stuck. I also try to explore new perspectives with them.

## Hard to define: mind, consciousness

Puzzling that psychotherapists, psychologists and psychiatrists, whose subject of study is the mind, haven't yet arrived at a unified definition of it. This is a significant problem faced by science: we still don't fully understand what consciousness is, and we don't even have a universal definition of the mind. One shortcoming of neuroscience has been trying to circumscribe the mind, and even human consciousness, to a byproduct of brain activity. It's postulated, for example, that the brain's thalamo-cortical system generates consciousness. However, the current state of the study of the brain doesn't explain certain phenomena, such as desire, ambition, or even parental love.

The John Templeton Foundation adjudicated a grant for The Cambridge New Directions in the Study of the Mind Project, directed by Tim Crane (July 2015–Dec 2017). At the Department of Philosophy and Theology of the John Templeton Foundation, a group of researchers led by Crane asked, "What kinds of new insights into, or what kind of progress in our understanding of, the mind and its power might be revealed by investigations informed by nonphysicalist or nonreductionist frameworks?" Crane and his team investigate the essence of the mind. They're clear on the idea that the nervous system provides, in some way, the mechanisms for thought and consciousness. But, according to Crane, the simple fact that something happens in the brain when you think doesn't explain what a thought is.

Doctor Larry Dossey, author of *One Mind: How Our Individual Mind is Part of a Greater Consciousness and Why it Matters* (Hay House, 2013), believes that our mind is not circumscribed to our body but that all individual minds are part of a collective consciousness. Dossey said, "Minds, rather, are nonlocal with respect to space and time. This means that the separateness of minds is an illusion, because individual minds cannot be put in a box (or brain) and walled off from one another."

"The human mind is, in a very real sense, much bigger and more expansive than the skull that we imagine (in our wrongly limiting way) to house it," neurobiologist Daniel Siegel says.[67] "It is an emergent, self-organizing, embodied and relational process that regulates the flow of energy and information. The brain and its whole body are the embodied mechanism of that flow, and the mind is the self-organizing process that regulates that flow." According to Siegel, "what we refer to as 'mind' emerges at the interface of interpersonal experience and the structure and function of the brain."

Physicist Fritjof Capra said,

> In the Santiago theory (of cognition) the relationship between mind and brain is simple and clear. Descartes' characterization of mind as the 'thinking thing' (*res cogitans*) is finally abandoned. Mind is not a thing but a process—the process of cognition, which is identified with the process of life. The brain is a specific structure through which this process operates. The relationship between mind and brain, therefore, is one between process and structure.

In his book *Rodolfo Llinás: La Pregunta Difícil* (*Rodolfo Llinás: The Difficult Question*, Aguilar, 2017), Pablo Correa transcribes a

[67] D. Siegel, "The Science of Consciousness." *Psychotherapy Networker*, May/June 2017.

conversation where the renown Colombian neurologist Llinás tries to answer a question beyond his understanding:

> "I have cornered the problem," he says energetically, with the hard look of a boxer analyzing his rival. "What are qualia?[68] Consciousness? Tell me." And he raises his voice. "Movement, motor skills, I understand them very well. I understand sensitivity well. Memory is a joke. Love. This is all BS. They are ridiculously simple things. But I do not understand how electrical activity translates into a sensation. The problem of qualia is a f… problem," he continues. "And it's going to be hard to understand because there's something we do not know yet. There is a large property that we do not understand. It's like trying to understand the sound that a bell generates, without understanding the vibrational resonance of the metal. It's a precious problem," he sighs. Then he looks pensive. In silence, he pauses to accept the fact that he is already 80 years old, that there is little time left and that he still does not have the answer he has pursued all his life. Then he says, "If someone told me, I will explain how consciousness works, but then I'll kill you, I would reply, perfect!"

Trying to understand the source of suffering, Buddha's followers have spent hundreds of years studying the mind. They know that, when not trained, the mind can become chaotic. According to Buddha, "All that we are is the result of what we have thought. The mind is everything. What we think we become."

Western cultures many attempts to define the mind range from describing it as the product of an individual's brain activity to understanding it as part of a collective consciousness. Psychoanalyst Carl Jung[69] believed a collective unconscious would explain our subconscious beliefs, our instincts and even spirituality.

After more than twenty-five years as a psychotherapist and a Reiki teacher, I see body and mind as indivisible, definitely not the mind as just a product of brain activity.

There is a popular belief that mental symptoms are the product of a chemical imbalance affecting brain function, and yet, no medication has ever been invented that could change brain circuitry. Neither Ativan nor Prozac nor Zoloft could unquestionably change the impact of the trauma you suffered, cure your anxiety, or permanently lift your depression.

---

[68] "The introspectively accessible, phenomenal aspects of our mental lives." *(Wikipedia)*

[69] C. G. Jung, *The Archetypes and the Collective Unconscious.* (Princeton: Princeton University Press, 1969).

Pharmaceuticals do have temporary effects on the availability of neurotransmitters in the brain, but unwanted and serious side effects should make us think twice before taking them. I'm not saying there aren't cases in which pharmaceuticals might be indispensable. But identifying and addressing the root cause of the symptoms promises a long-term change, and learning mindful emotional regulation seems significantly more promising than pharmaceuticals.

Yes, pills seem to help modulate mood and relieve symptoms. Pharmaceuticals can influence the levels of neurotransmitters available in the circuits associated with motivation, mood, memory, and stress responses and so they could alter your brain function, but the way these pharmaceuticals work is not very different from how alcohol, refined sugar, or recreational drugs affect the brain.

My point is that knowing your own body well, having an understanding on how it works, and making a commitment to changing your lifestyle and mental habits, are vital tools in this process of maintaining and recovering mental health.

A problematic aspect of diagnoses (not only mental health diagnoses) is that, most of the time, they lead to drug prescriptions that alter the production or binding of neurotransmitters and hormones. Many medications are potentially addictive, and, not being specific to the symptom, they produce undesirable side effects and long-term dependencies (see the opioid crisis). The multimillion-dollar pharmaceutical industry keeps focusing on genetic and biochemical research that could validate the need for the continued use of their products. Science and politicians are finally beginning to corroborate and understand the potentially irreparable damage the long-term use of pharmaceuticals can cause to the body.

If we don't yet have pills to treat a heartbreak, alleviate pain caused by disappointment, or even improve our capacity to feel compassion, it's because love is a much more complex matter than simple biology.

Let's now explore which parts of the brain are related to emotional regulation, how stress complicates our relationships, which circuits are activated when we feel empathy and compassion, and how the search for pleasure and reward also has its neurological substratum.

But when discussing the nervous system, let's consider what we already said about the mind: it's important not to reduce emotions, feelings, or thoughts to simple byproducts of neuronal activity. Just as my voice is not in the waves that transmit it from my phone to yours, and the people I watch on television are not in the cords that feed electricity to my devices or inside the TV box, in the same way, our thoughts and emotions are not inside our brain, our neurons, their cellular extensions (axons and dendrites) or in the molecules that transmit information (neurotransmitter peptides). These are all merely vehicles of a mysterious process. Therefore, when trying to express these connections, we need to bear in mind that words usually fall short when trying to explain such complicated concepts.

With luck, after reading, you'll be motivated to explore this intriguing topic further. Perhaps you'll agree with my assertion that we are not only biologically equipped for love but have the option of training our brains and our minds to promote empathy and regulate emotions. The idea that we have choices is empowering. We could choose to respond with empathy and love, instead of being driven by fear, reacting aggressively, competing, or developing addictive behaviors.

## Know Thyself

Jack Feldman, a neurobiology professor at UCLA's Brain Research Institute, made an interesting discovery on neonate mice: a small cluster of neurons sitting deep in the brainstem functions as a pacemaker for breathing. It's now known as the pre-Bötzinger complex. The publication of Feldman's findings in 1991 inspired further research, through which a similar structure has been identified in humans.

Mark Krasnow and Kevin Yackle, from the University of Standford,[70] led research on how this set of neurons affects not only breathing but emotional states and alertness. When they removed the neurons from the pre-Bötzinger complex in rats, they found that it was connected to a nearby circuit, the locus coeruleus (LC), which is associated with the breathing rhythm and controls anxiety. This circuit is the primary site for norepinephrine production in the brain, and it's involved in, among other functions, vigilance and attention. The LC distributes norepinephrine through the anterior brain and the limbic system, and its projections reach the spinal cord, through which it intervenes in both somatic and visceral (autonomic) motor control. The study may have important implications for the management of stress, anxiety, depression, and trauma sequelae.

Our mothers knew it: when they saw us out of control, they advised us to take a deep breath or count to ten (which also slows breathing). Popular wisdom has registered it: there is a correlation between physiological functions, such as breathing, and the brain that has an

[70] S. Sheikhbahaei and J. C. Smith, "Breathing to Inspire and Arouse," *Science Magazine* (March 31, 2017).

impact on our well-being and, eventually, on our state of consciousness. Meditators and mindfulness practitioners have known about this connection for centuries. Qi Gong practitioners, meditators and yogis also place special emphasis on conscious breathing for achieving stress reduction and increased awareness.

The knowledge we gain of our multidimensional body (physical, emotional, mental, spiritual, social), the questions we formulate trying to understand the purpose of symptoms, the learning of techniques such as Pranayama, and the ability to be mindful make a significant difference in the way we react to stressful events or record the memory of a trauma. Knowledge, then, also affects our physical and emotional health. The way we breathe could have an impact on the quality of our responses to stress and our state of consciousness.

Knowing oneself leads to understanding one's own nature and the potential of one's body and the mind, while making us aware of our limitations. Without knowing what our body is capable of and what actions we can take to fuel its innate wisdom or regulate our emotions or change a distorted perception of the world around us, it's difficult to lead a truly fulfilling life. Therefore, let's try to learn a little more about brain centers associated with our neuro-emotional responses.

But first, let me introduce the late neurologist Paul McLean, who worked at Yale Medical School and the National Institute of Mental Health a few decades back. He was instrumental in developing what he called the theory of the triune brain, according to which the human brain maintained its archaic structures through the evolutionary process.

McLean postulated that our brains consisted of three different evolutionary layers, each one developing from the previous: the reptilian brain, the mammalian brain, and the neocortex. According to his theory, each layer operates pretty much as an independent brain, with its own special intelligence, subjectivity, memory, and its own sense of time and space. Still, the three brains are interconnected and work in harmony with each other—most of the time. He demonstrated that the neocortex, the most recently evolved brain structure (involved in higher functions such as sensory perception, motor commands, conscious thought, and language), doesn't necessarily dominate the other layers of the brain. The limbic system (mammalian brain), which regulates emotions, can override conscious functions under certain circumstances.

According to McLean, from our reptilian ancestors we have preserved what he called the reptilian brain (R-brain or R-complex), which responds automatically to threats or changes in the environment by controlling the body's vital functions, such as heart rate, breathing, and body temperature. He described the R-complex in 1960 as the oldest part of our brain. It includes the brainstem and the cerebellum. It keeps the body functioning and regulates various processes, including growth and regeneration of tissues. It commands autonomic functions and muscle balance. It commands habits but also such responses as aggression, territoriality, rituals, and our sense of hierarchy. It is active even during sleep and keeps ancestral memory (summarized from different sources).

A mammalian brain responds to emotions such as fear by inducing the fight-or-flight response, activating the hypothalamic-pituitary-adrenal axis. It was McLean who, in 1949, introduced the term "limbic system" to name what is now considered our emotional brain, because of the activity observed in this area during emotional experiences.

The most important structures of the limbic or mammalian brain includes the hypothalamus, the hippocampus, the amygdala, and the pituitary. It seems to have evolved millions of years ago in non-primate mammals. It's believed to be programed with the language of emotions and to direct our experience. It is the center where memories are kept and evoked. If stimulated, certain emotions are provoked (fear, joy, anger, pain, pleasure). It processes most of our olfactory sensations. The limbic system is involved in emotional experiences that are required for survival: feeding, reproduction, and parenting, including the formation of attachments. The ventral tegmental area (VTA), involved in motivation and a sense of reward, is also part of the limbic system.

In the process of evolution, we have also acquired a neocortex (our "intelligent brain"), which is not exclusive to humans but also found in primates and even in some birds, such as magpies. It seems to have appeared evolutionarily as a result of the need to cooperate, compete, forge alliances. And deceive!

The neocortex appeared, it is believed, about a hundred thousand years ago. The outer layer of the brain, occupying two-thirds of the cerebral hemispheres, seems to be where matter is somehow transformed into consciousness. It is associated with abstract thinking, mathematics, and interpretation of sensory stimuli. It also regulates adaptability and maturity, which allow us to respond in a rational way to inherited or acquired patterns of behavior.

Other researchers have revised and enriched McLean's theory, but the basic idea that the human brain is organized in different layers, with functions that go from the most automatic and instinctive to the most elaborate and conscious, is still embraced by many psychiatrists and some neuroscientists.

Note that there is an interesting parallelism between the theory of McLean and that of Sigmund Freud, who believed the mind operates at three different levels: unconscious, preconscious, and conscious. Some esoteric traditions have also considered three different planes of consciousness. In the nineteenth century, the spiritual master Gurdjieff, for example, said that human beings had a brain for the body (corresponding to the automatic, reptilian brain), one for the soul (emotions, corresponding to the limbic system), and one for the spirit (consciousness, which would correspond to the neocortex).

To simplify, I might use Daniel Siegel's term "upstairs brain" to refer to the neocortex, particularly frontal cortex, associated with consciousness, and "downstairs brain" to refer to the limbic system and even the reptilian brain, which regulate our more automatic, instinctual responses. Our "upstairs brain" provides feedback to our "downstairs brain," letting us know if we're in a safe space. This is how instead of experiencing fear, people can tune the brain to a state that facilitates enjoyment or excitement.

## How do we react to the perception of danger?

When danger is imminent, nothing is more relevant to our multidimensional body than to keep us safe. Thus, the emotions induced by the perception of a threat will be dominant, and the physiological changes that prepare the body to face or avoid danger take precedence over any other aspects of our bodily functions.

The evolution of brain structures has allowed the human brain to design increasingly sophisticated circuits to respond to challenges presented by the environment. Moreover, most of the threats we face today can't compare to the perils our ancestors faced a hundred thousand years ago. Most of us don't live among tigers, serpents, or coyotes anymore. And if we do, we're very likely dwelling in sturdy houses that keep us safe. Today's dangers are of a different order, including nuclear threats and terrorism. On a more personal scale, digital technologies might eventually kill our jobs as they take on more of the attributes we've considered uniquely human, and artificial intelligence and robotics are inevitably becoming part of our lives.

Our survival in the wild (including the "urban wild") has depended, first of all, on the ability of our bodies to instinctively respond to danger and, secondly, on the possibility of meeting our basic needs, for which we have developed other brain mechanisms that motivate us to seek rewards (pleasurable experiences) and get them at a minimal cost. Experiencing pleasure guarantees the repetition of desirable behaviors that grant the survival of the individual or the species. There's less talk about how we're also hardwired to associate and cooperate with others, a response that increases the odds of both protecting ourselves and meeting our basic needs.

As living organisms evolved into greater complexity, they required additional energy, and their bodies demanded more nutrients. To gain access to these nutrients, they needed to develop ways to mobilize, which led to developing a nervous system that could coordinate movement. Because a moving body needs to make decisions about the direction in which to move, and when and how to do it, the brain got even more complex. At a certain moment, an organism needs to decide if it leaves its safe place to go in search of food. In the process of solving this dilemma, the brain developed new circuits that allowed for satisfying needs by obtaining pleasure, which became the motivation for most animal behavior. In vertebrates, this translated into an elaborate reward system mediated by the production of dopamine and serotonin.

In both the human brain and that of other mammals, several structures make up those reward circuits: the mesolimbic dopaminergic system, the amygdala, the hippocampus, the basal ganglia, the olfactory system, and the eyes. A newer structure in this circuit, the neocortex, became prominent in primates and the *Homo* genus. In humans, it covers the prefrontal and frontal lobes. The prefrontal cortex is involved in logical processes and decision making, and by evolving to access memories, it became capable of anticipating the future. It's interesting to note that, among primates, the larger their social groups, the larger the size of their neocortex, which is why this development is believed to correlate with social needs and especially cooperation.

In humans, the physiological and emotional responses to an event perceived as a threat are not limited to fight-or-flight, a mechanism described by Walter B. Cannon at the beginning of the twentieth century and commonly known as the primary response to stress.

Actually, we have not one but three different systems that regulate emotions and with which we respond to stress. In *The Compassionate Mind: A New Approach to Life's Challenges* (Harbinger, 2009), English psychologist Paul Gilbert describe them as:

a. A threat system that detects danger and initiates protective measures (fight-or-flight)

b. A drive system that motivates us to seek resources necessary for survival

c. A soothing system that promotes affiliation, comfort, and safety.

The first system tries to regulate our fear of destruction. The second is aroused when facing the fear of being eliminated by a rival and motivates us to compete, and the third (aroused by stressful external circumstances and loneliness) reduces our fear by eliciting the feeling of empathy and the formation of alliances.

Historically, stress studies have focused on the functions of the hypothalamic-pituitary-adrenal axis (HPA) and the secretion of cortisol in the adrenal cortex and catecholamines, such as epinephrine and norepinephrine (or adrenaline and noradrenaline), in the adrenal medulla (inner part of the adrenal gland) and the locus coeruleus.

A dopaminergic mesolimbic pathway has traditionally been associated with pleasure and reward,[71] but over the past three decades, new evidence has arisen that these dopamine-producing areas (dopaminergic neurons in the ventral tegmental area and the nucleus accumbens) are a second physiological response to stress that participates in actions aimed at reducing hunger, thirst, or sexual tension. Researchers have also observed that animals release dopamine in response to aversive stressful stimuli, such as separation of mother and infant. In humans, cortisol seems to facilitate the firing of dopamine neurons.

This second system, also involving the production of serotonin, is very active in people motivated to compete and work hard. Dopamine acts as a neurotransmitter in the brain (associated mainly with behaviors motivated by a reward) and peripherally as a molecular messenger, mostly at the level of blood vessels, inhibiting the production of noradrenaline. Stressful events, such as working against the clock or participating in a sports championship, increase production of dopamine, creating ideal physiological conditions to facilitate the person's completion of a job or participation in competition against rivals.[72] Chronic stress can lead to depletion of dopamine and can compromise blood flow to the brain (affecting memory, for example).

Shelley E. Taylor and colleagues at UCLA, described a third response to stress in 2000.[73] Apparently more developed in women and members of certain indigenous communities, this response is primarily

---

[71] B. C. Trainor, "Stress Responses and the Mesolimbic Dopamine System: Social Contexts and Sex Difference," *Hormone Behavior* 60 (5) (Nov. 2010) 457–469.

[72] E. N. Holly and K. A. Miczek, "Ventral Tegmental Area Dopamine Revisited: Effects of Acute and Repeated Stress, *Psychopharmacology* Volume 233, Issue 2 (Jan. 2016): 163–186.

[73] S. E. Taylor, "Biobehavioral Responses to Stress in Females: Tend-and-Befriend, Not Fight-or-Flight," *Psychological Review* Vol. 107, No. 3 (2000): 411–429.

involved in our protection of children and the most vulnerable people in the community. This physiological response that stimulates cooperation, alliances, mutual support, and the creation and maintenance of social networks is now known as the tend-and-befriend system.

Prolonged activation of the fight-or-flight response leads to adverse effects in the body, due to the secretion of adrenaline and, especially, cortisol. In the long term, stress compromises the body's metabolism, the cardiovascular system, and the immune system. Chronic psychosocial stress can also lead to symptoms of anxiety, depression, negativity, and hyperactivity. It's easy to figure out how difficult it becomes to love oneself or others when our emotional and physical health are compromised or when the presence of stressors promotes automatic responses of aggression or avoidance.

Dopamine is related to the intense emotions and focused attention we experience in the initial stages of falling in love and to the pleasure found in sexual activity. An increase in serotonin in the brain (for example, when taking antidepressants or eating a diet rich in refined sugar) tends to suppress the dopaminergic system and compromise the ability to experience love.

A very different response occurs when the third system is activated, because the hormone oxytocin, modulator of this system, counteracts the effect of cortisol and is involved in the formation of bonds between people. Positive social interactions then can counteract anxiety and reduce the detrimental effects of stress.

Both men and women release oxytocin but, interestingly, the female hormones—estrogens—make the receptors in the ventromedial nucleus of the hypothalamus more receptive to oxytocin.

## Emotion regulation and stress response

For us humans, our thoughts, beliefs, lifestyles, learning processes, jobs, careers, choices of partners are ultimately motivated by the need to survive as individuals and as a species. The loss of our support systems, loneliness, lack of access to healthcare, unemployment, or the inability to pay our rent, mortgage, or credit card bills are all modern threats to our survival. And any threat rouses fear and activates our stress response systems. Interestingly, the same brain areas activated by stress fire when you make a conscious effort to regulate emotions.

We established that when human beings face situations that threaten their well-being, health, integrity, and even their social status, they activate physiological responses that help them cope with the perceived threat. Many people are already familiar with the fight-or-flight response. Those who have learned about addictions would have a good knowledge of the pleasure and reward system (not always recognized as a stress response), but the affiliative response has only been studied in the past couple of decades.

Let's recapitulate. Our main three known emotion regulation or stress responses are:

1. Fight-or-flight response

2. Action motivated by the search for pleasure or rewards—a response that stimulates competition (survival could depend on eliminating rivals)

3. Looking for affiliation, taking caring care of others, letting others care for us, forming alliances, and making friends.

Interacting and counterbalancing each other, these three responses correlate with our experience of different emotions and mental states. Specific brain circuits and physiological processes are involved in the function of each system that guarantees not only our individual survival but that of our species.

They are then, adaptive physiological responses to tensions and perceived threats in the environment.

And now, let's dive a little deeper.

## First response—Fight-or-flight

The reader must have experienced a few scares. Perhaps you've witnessed a robbery or a car accident or watched scary thrillers or alarming news on TV. At first, the brain doesn't seem to distinguish real from imaginary. So, no matter what the alarm is, you might have noticed that when you feel frightened, your heart beats faster, and your breathing hastens, and also that your attention focuses on the event that just happened or that you're anticipating. You must also have noted that your skin stays warm and damp and your heart keeps pounding even minutes after you confirmed you were safe. In a matter of seconds, as soon as the body perceived the threat, it created the physiological conditions to enable you to either flee from danger, attack an aggressor, or defend yourself. That's a normal reaction to any perception of imminent danger.

Walter Cannon coined the term fight-or-flight in 1932 to describe this response characterized by the secretion of adrenaline, noradrenaline (also known as epinephrine and norepinephrine), and cortisol. He understood it was an autonomic nervous system mechanism that allowed animals to respond to danger.

In 1946, Dr. Hans Selye published his own valuable contribution to the understanding of stress and dedicated it to Cannon. Selye described the "general adaptation syndrome," with which an organism responds to prolonged stress. Although today the practical value of his first contribution is debated, it served as a conceptual framework for Selye and his colleagues to develop a neuro-hormonal model of stress that demonstrated the role of stress—through the alteration of the function of the pituitary and adrenal glands—in the etiology of many chronic diseases, such as hypertension, arthritis, gastric ulcers, asthma, and cancer.

After a threat is perceived, the hypothalamic-pituitary-adrenal axis is activated. Any situation where physical danger seems imminent activates the response—an accident, an assault, a sudden noise—but it can also be activated by an argument with a loved one or by someone verbally and emotionally abusing us. When the amygdala, in communication with the hypothalamus (each a part of the brain's limbic system), is activated, the sympathetic system starts an automatic response to prepare the body to either fight or flee. On occasion, a stimulus overwhelms our coping capacities, and we freeze in fear, instance in which the parasympathetic system hits the brake on the motor system.

In general, stressors engage a whole network of limbic structures that include the insula, amygdala, and anterior cingulate gyrus, as well as thalamic, hypothalamic, and lower brainstem sites. Feelings of fear, anger, anxiety, or disgust typically accompany this response to stress. We can see the same responses in animals. Rabbits, for example, tend to freeze when they feel us approaching, fawns freeze or flee, snakes flee or attack, tigers, most frequently, attack.

The adrenaline and cortisol secreted by the adrenal glands facilitates the redistribution of blood circulation to the body parts that would be involved in fleeing or fighting, mainly muscles, heart, and lungs. While fear or the urge to attack prevail, the areas of our cerebral cortex responsible for processing emotions and using logical responses shut down, which explains, among other things, why it's useless to argue with a person who's overcome by anger or fear. As our brains become increasingly sophisticated, humans are able to learn to focus attention, subdue aggressive impulses, and offset the need to flee from a situation initially perceived as threatening but evaluated as safe.

Now let's talk about the brain structures involved in this first stress response. However, bear in mind that stress responses comprise an interface between perception (what you see, what you hear), a fast evaluation of sensory input, and a subsequent behavior.

The limbic system is in charge of processing the fight-or-flight response and storing key bits of information after a threat is gone. We need to learn from danger. The hypothalamic-pituitary-adrenal axis is a complex set of neuroendocrine interactions between the brain and the adrenal glands that modulate the primary response to stress, and the fact that it also regulates other vital processes, such as digestion, the immune system, metabolism, and sexual conduct, might explain physiological symptoms associated with stress.

## In more detail

Perception of danger through our senses activates the amygdala, the hypothalamus, and some areas of the left frontal cortex. The experience of fear involves a) a gut feeling (signals that the amygdala is activated), b) an urge associated with survival (hypothalamus is activated), and c) some thinking (frontal cortex is active).

1. **The amygdala**, on both sides of the frontal part of the temporal lobe, is associated with the processing of emotions (such as anger, fear, sadness) and the controlling of aggression and anxiety, as well as satisfaction. It's deemed responsible for the acquisition and expression of conditioned fear responses. When a treat is perceived, the processes that lead the body to either attack, flee, or freeze take place in the amygdala, but the decisions are ultimately processed in the cortex. However, recent research seems to indicate that danger signals originating within the body (damaged tissues, infections, inflammation), a primary form of fear, may not be processed through the amygdala.

2. **The hippocampus**, located beneath the amygdala, has a role in the interpretation of the emotional significance of events and the storage of memories that can help the body discern what actions are required to avoid or confront a threat.

Both the amygdala and the hippocampus control the activity of the hypothalamus-pituitary-adrenal axis, which determines the secretion of the so-called stress hormones (noradrenaline and cortisol).

3. **The hypothalamus**, located above the amygdala, bridges the autonomic nervous system and the endocrine system (via the pituitary gland). By regulating the functions of the sympathetic and parasympathetic nervous systems to maintain homeostasis, the hypothalamus sets in motion and coordinates the essential bodily processes needed not only for our survival but to ensure that the body can adapt to changes in environmental conditions or variations in the

functioning of the body. The hypothalamus communicates with the prefrontal cortex, which contributes to the regulation of emotions and which is in charge of processes associated with foreseeing the future and planning actions.

4. **The pituitary**, a tiny organ located at the base of the brain, considered the master gland of the body, secretes hormones that regulate the function of the other endocrine glands.

5. **The thalamus** relays incoming sensory information to areas in the cerebral cortex to be processed. It not only provides the cortex with sensory input but facilitates communication in the cortex between sensory, motor, and cognitive functions. The thalamus is involved in the regulation of alert states, which is essential when encountering a threat.

All of the above-mentioned regions are involved in the primary stress response. They also intersect with neuronal circuits responsible for memory storage and reward-seeking behavior.

Based on previous experiences and the anticipation of an outcome, the brain "chooses" how to react in the face of a threat. When stress is intense or becomes chronic, memory could be impaired, and therefore the capacity of the body to respond will also be compromised.

6. Selecting the appropriate response to a threat is a task of **the ventromedial prefrontal cortex (vmPFC)**, a region of the brain located in the most anterior part of the frontal lobe. It seems to be involved in higher cognitive processes, planning, personality formation, and acceptable social behavior. Some researchers believe that the vmPFC is in charge of integrating information, and therefore would be at the center of decision-making processes. Some scholars have also observed that this part of the cortex could play a role in the recovery of long-term memory.

Let's dedicate a special section to the cortex.

## Prefrontal cortex

There are three main areas of the prefrontal cortex: lateral (lPFC), medial (mPFC), and orbitofrontal (OFC).

The ventromedial prefrontal cortex (vmPFC) is a part of the orbitofrontal cortex, located in the bottom part of the frontal lobe. It's involved in the processing of threats and fear. It's extensively and reciprocally interconnected with the amygdala, the temporal lobe, and the anterior, medial, and ventral regions of the striatum. It seems to be involved in decision-making dependent on somatic (emotional) signals coming from the amygdala, the learning processes that include an affective involvement, a propensity for risk, and impulsivity, and it plays a key role in the regulation of emotion, for which it has been dubbed the dopaminergic circuit's brake.

It's extensively and reciprocally interconnected with the amygdala, the temporal lobe, and the anterior, medial, and ventral regions of the striatum. It seems to be involved in decision-making dependent on somatic (emotional) signals coming from the amygdala, the learning processes that include an affective involvement, a propensity for risk, and impulsivity, and it plays a key role in the regulation of emotion, for which it has been called the dopaminergic circuit's brake.

Researchers have observed that increased activity in the vmPFC is associated with decreased activity in the amygdala, which has led them to conclude that it plays a role in the extinction of conditioned fear responses.

At the University of Lethbridge, in Canada, David R. Euston and colleagues[74] proposed that the function of the prefrontal cortex is to learn associations between context, place, and event, and then issue the needed adaptive responses, especially emotional responses. That's why the prefrontal cortex is thought to be involved in both memory functions and decision-making, since it is crucial to be able to recover, from the memory archive, those actions that have already been effective in certain places and moments.

One of the most important functions of the neocortex, and in particular of the medial prefrontal cortex (mPC), is the integration and processing of sensory impressions to build memory, perception, and cognitive activities.

Once information from the environment is captured by the senses, it goes through the amygdala, which responds instinctively to any sense of danger. From there, the information goes to the hippocampus, responsible for the processing of long-term memory. The mPC computes the information to allow the brain to choose the most appropriate response to an experience to avoid danger and solve problems. Researchers concur that activation of this part of the cortex is associated with logical and conscious thinking.

It's worth mentioning that repetitive stress leads to a remodeling of neurons and neural connections in the amygdala, the hippocampus and the prefrontal cortex. These changes, the result of chronic stress, include a shrinkage of neuronal dendrites and a reduction in the density of dendritic spines in the neurons of the mPC, pyramidal neurons, and neurons of the dentate gyrus.

[74] D. R. Euston, A. J. Gruber, and B. L. McNaughton, "The Role of Medial Prefrontal Cortex in Memory and Decision Making," *Neuron* 76(6) (Dec. 20, 2012): 1057–1070.

Chronic stress also decreases the production of new neurons (neurogenesis) and affects the population of neurons in the dentate gyrus (part of the hippocampus). However, the dendritic branching in the orbitofrontal cortex (involved in the decision-making process) increases as a result of chronic stress. For the most part, these stress-induced changes in the hippocampus and the medial prefrontal cortex are reversible over time, at least in the animal models that have been researched so far. Because these regions of the brain are interconnected, structural remodeling in one region is likely to influence the functions of other regions of the brain.

Some scholars agree that the mPC participates as a mediator in the decision-making process; other researchers believe it's involved in the recovery of long-term memory.

There's a wealth of information online on these topics, should you want to know more about the first response to stress. Explore for example, https://courses.lumenlearning.com.

## Second response: Incentives and rewards

Mediated by the secretion of dopamine, this second response is critical for psychological motivation and physical movement and allows for the association of pleasure with activities related to survival. Dopamine is involved in the regulation of certain cognitive skills essential to human language and thought, and it plays a role in sociability and social anxiety. A drop in the dopamine levels is associated with an observable decrease in the desire to work and the ability to experience pleasure. Serotonin, endogenous opiates, and GABA modulate dopamine levels.

Even what would be known, in a psychiatry scale, as "negative symptoms" of schizophrenia (social withdrawal, apathy) have been linked to a dopamine deficiency, while dopamine function has been found to be increased in correlation with schizophrenia's "positive symptoms" (hallucinations, delusions, racing thoughts).

This more sophisticated system, also known as the drive system, is involved in the motivation to act and to look for resources, solutions, and results. Say, for example, that we feel we're in danger of losing our job (and therefore our livelihood). The threat can encourage us to notch up our qualifications, submit an innovative project, become indispensable at work, or acquire new skills. When activated, we feel compelled to succeed and to seek the resulting pleasure from achievement, which would reduce tension.

Like in the previous, this second system is also mediated by neurotransmitters (in this case, dopamine and serotonin). The corresponding neural circuits (also known as the mesolimbic dopamine pathway) comprise the following structures:

1. **The ventral tegmental area** (VTA), with neuronal projections to…
2. **The nucleus accumbens** in the ventral striatum,
3. The orbitofrontal cortex, and
4. The prefrontal cortex

The VTA, located in the midline of the midbrain, has dendritic projections toward the nucleus accumbens and the prefrontal cortex. It's an area rich in opioid receptors that, when stimulated by endorphin release in pleasurable situations, certain drugs or alcohol, stimulate the production of dopamine or inhibit its recapture. It plays an important role in cognition, motivation, and orgasms.

The VTA monitors the environment in search of situations that favor survival. When those situations are found, the VTA releases dopamine into the mesolimbic pathway toward the NA, and the person experiences the sensations of pleasure and satiety, or satisfaction.

Researchers noted that when the dopamine system is activated, people tend to be excited and more motivated to pursue their goals or seek rewards. However, recent research shows that dopamine levels in the nucleus accumbens are elevated both in response to stimuli related to reward and to aversive stimuli. This suggests that dopamine may be involved in the storage of information related to all kinds of stimuli in the environment that would invite us to repeat pleasant experiences and avoid the unpleasant ones.

## More about dopamine and serotonin

The most important neurotransmitter in the drive system, dopamine, is secreted in the ventral tegmental area (VTA) and the nucleus accumbens (NA), but also in the hypothalamus. It plays a role in executive functions, such as attention, exploration, and creative impulse. It's also associated with the intensity of our emotions. Dopamine is therefore involved in many aspects of our behavior, from the creation of incentives and the focus of our attention, to pleasant feelings that reinforce our behavior. Low levels of dopamine have been associated with the onset and development of Parkinson's disease.

Neurotransmitters are part of the body's internal communication system. They facilitate the transfer of electrical or chemical signals between nerve cells. Neurotransmitters are secreted by a neuron into its synapse (a space where neurons meet without touching) where they bind to specialized proteins called receptors, which exist on the surface of a neighboring neuron, to start transmitting signals. Another kind of neurotransmitters, known as transporters, remove an already used neurotransmitter from the synapse, either to prevent its further action or to allow it to be reutilized. Neuromodulators are those slow-acting neurotransmitters that are not reabsorbed or reutilized.

These slow-acting neurotransmitters (dopamine, serotonin, acetyl-choline, noradrenaline) promote cellular changes, enlarging synapses or promoting the formation of new ones. The more synapses, the easier to use a neural pathway. Slow-acting neurotransmitters are thus involved in processes such as memory and learning, repetitive actions, and addictions.

Whether a neurotransmitter is excitatory or inhibitory depends on the existence or absence of a certain receptor in the next-in-line neuron. Dopamine acts as both an excitatory and an inhibitory neurotransmitter. Serotonin is more often considered an inhibitor neurotransmitter. Both dopamine and serotonin are associated with changes in mood, memory, learning, concentration, sleep patterns, and motor function.

In a process that starts at birth, any pleasant activity stimulates the dopaminergic (dopamine producing) and serotoninergic (serotonin producing) circuits, conditioning us to learn and repeat certain behaviors. Science has, for example, identified foods that induce the secretion of dopamine (chocolate, cheese, avocado, apples, bananas) and foods that stimulate the production of serotonin (cheese, eggs, tofu, nuts, seeds, sugar and, again, chocolate), and we've learned to consume them (or crave them) not necessarily to alleviate hunger but to reduce stress. That's why they become comfort foods: they make us feel more at ease, at least temporarily.

The sensations of pleasure and well-being associated with situations and behaviors that increase dopamine levels eventually condition us to further seek and desire situations and actions that stimulate its secretion. If eating and having sex were not rewarded with pleasure, maybe we would neglect feeding ourselves, and maybe we'd feel no need to mate, both of which, of course, would be detrimental to the preservation of our species. However, many people also learn to stimulate the secretion of the "feel-good" neurotransmitter by using addictive drugs or repeatedly engaging in rewarding behaviors—such as playing video games, and sometimes participating in risky activities that flood the nucleus accumbens with dopamine. The problem is that when dopamine interacts with glutamate (another neurotransmitter), the reward learning system is hijacked. Addictions are therefore associated with a repeated spike in the release and eventual depletion of dopamine in the nucleus accumbens, The continuous use of substances that stimulate the production of dopamine, leads to habituation and physical dependence.

Apparently, dopamine plays a key role in the development of certain human cognitive functions, such as abstract thinking and working (short-term) memory, thanks to which we can store and manipulate the necessary information to be able to perform complex processes, including analyses, mathematical operations, and language comprehension.

Researchers have studied the correlation between diet, aggressiveness, and achievement-oriented behavior. Discussing nongenetic explanations for increased dopamine levels at a certain point in human evolution, in an article published by *Brain and Cognition* journal, Fred H. Previc[75] noted that a diet with an increased meat and shellfish consumption "led to greater supplies of dopamine precursors and conversion of them into dopamine." Although there are no definite conclusions about a cause-and-effect relationship, it's interesting to note that India—the cradle of yoga, nonviolence, and meditative practices—continues to be the country with the highest percentage of vegetarians (38 percent).

On the other hand, a correlation has been observed between increased levels of dopamine and more aggressive but less emotional behavior, including risky behaviors such as gambling, but especially behaviors aimed at climbing the social ladder, winning a championship, or getting a promotion.

Seo and colleagues[76] have found that "dysfunctional interactions between serotonin and dopamine systems in the prefrontal cortex may be an important mechanism underlying the link between impulsive aggression and its comorbid disorders."

Some psychiatric diagnoses have been associated with an excess of dopamine, particularly attention-deficit, bipolar, and obsessive-compulsive disorders, as well as autism and schizophrenia. A deficit of dopamine has also been observed in depression (although serotonin

---

[75] F. H. Previc, "Dopamine and the Origins of Human Intelligence," *Brain and Cognition* 41 (1999): 299–350.

[76] D. Seo, CJ.Patrick, and PJ. Kennealy. "Role of Serotonin and Dopamine System Interactions in the Neurobiology of Impulsive Aggression and its Comorbidity with other Clinical Disorders." *Aggression and Violent Behavior.* (Oct. 2013) 383-395. Retrieved online October 7, 2019: www.ncbi.nlm.nih.gov/pubmed/19802333/

deficits are more often associated with depression). However, there's no clear baseline to establish what levels were normal before a diagnosis was established, and the chicken-or-the-egg question looms: What was first, the neurotransmitter deficiency or the symptoms?

An increased production of dopamine could account for what some consider great achievements of humanity. However, greed, which is considered an important factor contributing to overproduction crises, is also associated with high levels of dopamine, leading some researchers to wonder if human behavior, particularly competitive actions, hasn't reached a point at which it's compromising our survival and that of the planet as well.

In *The Dopaminergic Mind in Human Evolution and History* (Cambridge University Press, 2009), Previc claims we've become a hyperdopaminergic society. Behaviors we described above as obstacles to experiencing love, seem mediated by a higher production of dopamine: the frenetic pace of life and the search for novelties, money, and fame; narcissism and the obsession with achievements; and the increase in highly competitive environments.

It's interesting to note that an individual might go in search of exciting competitive or risky activities, and then self-medicates, to calm down, with practices like mindfulness,[77] yoga, meditation, or a quick fix involving downers–substances that modulate dopamine production.

## Third response: tend-and-befriend

If you were walking alone through a dark and lonely alley at midnight and three masked individuals were coming at you, this would not be the best time to try and befriend strangers. The most advisable thing for you to do in such circumstances is to follow your instinct: move to the other side of the street, speed up your pace, get ready to run or scream for help if need be. Fight only if the situation and your Kung Fu skills merit it. However, if a hurricane or an earthquake just happened, go and offer or ask for help from your neighbors. A community can, with much higher success than an individual, overcome a calamity. This explains the significance of support groups and nonprofit organizations, especially in cases of natural disasters or war: people coming together to multiply resources.

[77] P. C. Broderick, "Mindfulness and Coping with Dysphoric Mood: Contrasts with Rumination and Distraction," *Cognitive Therapy and Research*, Vol. 29, No. 5, (Oct. 2005) 501-510 doi:10.1007/s10608-005-388.

The tend-and-befriend system can be observed in all mammals. It stems from an instinct to care for others and an impulse to form alliances and to congregate in search of support in times of crisis. This response is believed to have appeared about 120 million years ago, with the evolution of the mammals, as part of a physiological system in charge of bringing about sedation.

This third system is associated with our innate need to connect, to socialize, to seek alliances, to approach others to relieve tension, as in the following vignette.

*While he spends hours at the computer seeking solutions to his business-related troubles, his wife reacts to his aloofness by socializing more than ever. He's irritable and not eating or sleeping well. She understands and feels compassion because she knows he's facing the closing of a business it took him thirty years to build, but she's also angry and resentful because they're no longer the couple they used to be. She feels she's been shoved to the bottom of his priorities. She knows he's going through hell because of mistakes he made and that he's hurt by the fact that his children (from a former marriage) are not helping. She wouldn't know how to go through this crisis without the support of her friends. She doesn't even have to discuss with them what's going on at home. It's just that company makes her feel so much better. Socializing is the best way for her to cope with stress.*

When we seek to vent emotions after a tough day, when we join efforts during a crisis, when we respond in solidarity to help the victims of a disaster, we're using this affiliation response, which induces feelings of security, connection, and contentment. We use this response because it makes us feel better—relieves tension. It facilitates the regulation of emotions through an empathic and compassionate expression.

We've observed this response in action in romantic relationships or between parents and children, especially in times of turmoil. We feel relieved in the presence of those who are truly present with us, who show us authentic concern and affection. Scientific evidence shows us that this response could be learned and cultivated if we put our will into it. It's a biological response that allows us to feel empathy, compassion, and, if the bond lasts, love.

We owe thanks to Shelley Taylor and colleagues for the development of the tend-and-befriend theory, which maintains that "under conditions of threat, tending to offspring and affiliating with others (befriending) are common responses in humans." They argued that there was an evolutionary reason behind the development of this response primarily—but not exclusively—in women: the differing roles and degrees of participation of men and women in parenting. While men were out there hunting, women needed to protect themselves and their children, especially during pregnancy, lactating, and raising young children.

Taylor says,[78]

> Of interest is the fact that the causes of death that largely account for men's early mortality are those related to the fight-or-flight response, namely, aggressive responses to stress, withdrawal in the form of substance abuse, and coronary artery disease, the risk for which is exacerbated by frequent or recurrent stress exposure. By contrast, women more reliably turn to their social contacts in times of stress, responses that are...protective of health and longevity.

A tend-and-befriend response is also observed in tribal communities, especially those that have been minimally exposed to the stressors of modern life.

Researchers found that women regularly create, maintain, and use the support of social groups to protect themselves emotionally and manage stressful situations. Women seek social interaction to alleviate tension, which increases the secretion of oxytocin, endorphins, and estrogens.

On the other hand, men seem to prefer handling their tension through being by themselves or participating in strenuous activities (working more hours or spending more time working out) and, less often, doing some male bonding. Their responses to stress are mediated by adrenaline, cortisol, and dopamine, but also testosterone, which has been associated with competitive and aggressive behaviors.

While emotional, defensive, and moderate confrontations seem to be more prevalent in females, males are more focused on aggressively seeking resolution of stressful situations. In general, women handle stressful situations by seeking support instead of fighting or fleeing.

Based on the hypothesis that estrogens attenuate the effects of stress hormones, Tricia Adjei[79] and colleagues studied sex differences in electrocardiograms of subjects exposed to stress. Their study suggested for the first time that "the conventionally cited cardiac changes,

[78] S. Taylor, "Biobehavioral Responses to Stress in Females: Tend-and-Befriend, Not Fight-or-Flight," *Psychological Review* Vol. 107, No. 3 (Jul. 2000): 411–429.

[79] T. Adjei, J. Xue, and D. P. Mandic, "The Female Heart: Sex Differences in the Dynamics of ECG in Response to Stress," *Frontiers in Physiology*. 9:1616. (Nov. 28, 2018). Retrieved online June 10, 2019.

attributed to the fight-or-flight stress response, are not universally applicable to females. Instead, this pilot study provides an alternative interpretation of cardiac responses to stress in females, which indicates a closer alignment to the evolutionary tend-and-befriend response."

| SYSTEM | RESPONSE | NEURO-MODULATOR | BRAIN AREA |
|---|---|---|---|
| **Fight-or-flight** | **Active:** fight-or-flight<br><br>**Inhibited:** Freeze. No response | Adrenaline<br>Noradrenaline<br>Cortisol | **Active:** amygdala, hippocampus, and medial prefrontal cortex |
| **Incentives and reward—mesolimbic dopaminergic path** | Competition, excitation, motivation in search for solutions, pleasure or reward. Need to eliminate competitor who threatens survival | Dopamine<br>Serotonin<br>Testosterone | Nucleus accumbens, VTA, orbitofrontal cortex, and medial prefrontal cortex |
| **Tend-and-befriend** | Alliances and mutual support to grant survival | Oxytocin<br>Estrogens<br>Endorphins | Hypothalamus pituitary, cortex |

***The chart summarizes anatomical substrates, hormones, and neurotransmitters involved in stress responses, and the characteristic response of each of the three systems described above.***

## Oxytocin, the bonding hormone

Oxytocin (OT), the neuromodulating hormone associated with this tend-and-befriend system, is mainly produced in the hypothalamus, from where it is secreted into the bloodstream through the posterior pituitary gland or other parts of the brain and spinal cord. The secretion of oxytocin depends on the electrical activity of the neurons in the hypothalamus. Estrogens seem to stimulate the release of oxytocin in the hypothalamus but also to facilitate the binding of oxytocin to the membrane receptors of the neurons in the amygdala and the ventral hypothalamus.

OT is synthesized not only in the brain but throughout the body, including in the heart, thymus, gastrointestinal tract, and reproductive organs. Receptor cells for oxytocin are also found in the brain and other parts of the body such as ovaries, uterus, mammary glands, testis, liver, adrenals, and fat, and its expression changes over the course of our development.

Something especially interesting about oxytocin is that besides being a hormone (a chemical that stimulates specific cells or tissues in the body to perform specific functions), it's also a neurotransmitter (which facilitates communication between neurons). OT is sometimes known as the "love hormone" because, among other things, it's found in high concentrations in the blood of people who are starting a romantic relationship. But OT is not the only neuromodulator involved in love. When two people are attracted to each other, the brain first produces phenylethylamine (an alkaloid that's also found in chocolate), and then it secretes norepinephrine and dopamine. Oxytocin, endorphins, and serotonin are also produced in subsequent stages of a relationship.

For many decades after it was discovered, the only thing we knew about OT was that, at the time of labor, it was responsible for the intensity of the mother's uterine contractions and the dilation of the uterine cervix. OT is therefore often used to induce or help women speed up the time of labor. Later on, it was observed that the mother's brain also secreted OT while breastfeeding, triggering uterine contractions that contributed to the uterus shrinking back to its pre-pregnancy size.

Nowadays, we also know that OT has a role in the modulation of parenting behaviors, the formation of bonds between mother and child (both secrete it) and lowering maternal stress. It plays an important role in the establishment of bonds other than those between mother and child and contributes to the lowering of both our heart rate and cortisol levels. OT is released during orgasms, and even when we hug a friend.

Let me add a brief note about other recent findings.

Strathearn and colleagues[80] scanned first-time mothers while they were looking at a picture of their baby smiling. They found that dopaminergic (reward system) areas in the brain lit up. Researchers have

[80] L. Strathearn, J. Li, P. Fonagy, S. McClure, C. Bracero, K. Brartley, and P. Read Montague, "What's in a Smile? Maternal Brain Responses to Infant Facial Cues," *Pediatrics* 122(1) (July 2008): 40–51.

wondered how this activation of the reward system contributes to the establishment of a bond between mother and child. The smile of the baby, they think, would play as a survival strategy, starting the socialization process for the baby. A notion already existed that babies' soft features triggered feelings of tenderness, which served the purpose of guaranteeing their care.

Although OT was discovered more than a century ago, the interest in oxytocin has recently increased as a result of studies showing it has a role in social behavior. The precise mechanisms are not yet fully understood, but the release of OT seems to have an impact on dopaminergic activity, by stimulating not only the reward system but the expression of other affiliative behaviors, including those aimed at protecting a relationship, such as jealousy (competing for a partner) and envy (competition for the care of your partner).

Recent studies[81] suggest that oxytocin promotes prosocial behaviors and helps reduce social anxiety by helping people bond with each other.

These new discoveries bring hope that OT could be used in the treatment of certain conditions where social functioning is impaired. For example, after inhaling the hormone, children diagnosed with autism improved their prosocial behavior for a short time (they were more attentive to social cues), which has led to the funding of larger studies seeking confirmation.[82] Research has also suggested that oxytocin might have positive outcomes in the treatment of addictions (for example, to methamphetamine).[83]

Since Michael Kosfeld and colleagues published "Oxytocin Increases Trust in Humans" in 2005 in *Nature* magazine,[84] several other studies seem to reaffirm that oxytocin influences the degree of trust felt toward others. However, in a critical review of existing research on the

---

[81] C. Jones et al.'s "Oxytocin and Social Functioning," *Dialogues in Clinical Neuroscience* 19(2) (June 2017): 193–201.

[82] K. J. Parker, O. Oztan, R. A. Libove, D. Sumiyoshi, L. P. Jackson, D. S. Karhson, J. E. Summer, K. E. Hinman, K. S. Motonaga, J. M. Phillips, D. S. Carson, J. P. Garner, and A. Y. Hardan, "Intranasal Oxytocin Treatment For Social Deficits And Biomarkers Of Response In Children With Autism," *PNAS* (July 10, 2017).

[83] B. M. Cox, B. S. Bentzley, H. Regen-Tuero, R. E. C. M. See, M. Reichel, and G. Aston-Jones, "Oxytocin Acts in Nucleus Accumbens to Attenuate Methamphetamine Seeking and Demand," *Biological Psychology* 81(11) (June 1, 2017): 949–958.

[84] M. Kosfeld, M. Heinrichs, P. J. Zak, U. Fischbacher, and E. Fehr, "Oxytocin Increases Trust in Humans," *Nature* Vol. 435, No. 7042 (June 2, 2005): 673–676.

topic, published in 2015 in the *Perspectives in Psychologic Science*, Gideon Nave et al could find no conclusive evidence.

While more research is needed, here are a few encouraging findings about oxytocin's effects:

- There is evidence that OT contributes to the growth of the neocortex, and the maintenance of the blood supply to the cortex, which seems to be a factor contributing to encephalization (an increase in complexity and size of the brain).

- There may be a positive correlation between oxytocin, trustworthiness, and generosity.

- Men who secrete more OT might be more faithful, maybe because OT contributes to them seeing their partners as more attractive.

- OT seems to play a role in bonds established among people perceived of as belonging to the same group or race (ethnocentricity).

- OT may contribute to connect the brain circuits related to processing social information (faces, noises, smells) to the reward system.

Even if science has not yet corroborated all of the above findings, and if sometimes the studies have rendered contradictory results (such as associations between oxytocin and aggressive behavior, or with envy and gloating), there is increasing evidence that oxytocin facilitates prosocial behaviors and plays a role in social cognition in general. Researchers continue looking for oxytocin practical applications.

## One more

Before finishing this section, I felt it was necessary to mention the polyvagal theory, addressing a relationship between the central nervous system and visceral experiences. A very long bundle of fibers, the vagus nerve is part of our autonomic parasympathetic system. It originates at the top of the spinal cord (at the brainstem) and it transmits information about the body to the brain and plays a role in the regulation of the functions of the heart, lungs, muscles, digestive tract, metabolism, and other viscera.

When the vagus is activated, the body relaxes. Transcutaneous stimulation of the auricular branch of the vagus nerve is being used (especially in Europe) to treat certain disorders such as drug-resistant epilepsy. Such stimulation also seems to contribute to the regulation of stomach acidity, the production of digestive enzymes, and an increase in the contractions of the stomach to empty its contents toward the intestines.

The vagus also provides innervation to the intestinal tract and plays a role in the regulation of appetite and mood. There is evidence that the immune response, which regulates inflammatory processes, might be modulated by the vagus nerve.

In 1994, Stephen W. Porges, at the University of Indiana in Bloomington, proposed the polyvagal theory, linking the evolution of the autonomic nervous system to social behavior. He proposed the existence of a vagal circuit—related to structures located on the right side of the brainstem—which may intervene in the expression of emotions, working together with neurophysiological and neuroendocrine systems.

Whenever you've been under the effects of an adrenaline rush, you felt the palpitations of the heart, the accelerated breathing, the movement of the intestines, the warm and damp skin. Well, the activation—or stimulation—of the vagus nerve produces the exact opposite sensations, such as those experienced while listening to soft music, appreciating a sunset, meditating, receiving a Reiki session or a massage, or even when feeling grateful.

The vagus nerve seems to connect to the oxytocin neural pathways and would be related to our ability to respond to the sound expressions of pain and joy. Branches of the vagus go to specific muscles of the throat that intervene in the vocalization of sounds and thus, could play a role in verbal communication. All these characteristics have led several researchers to consider the vagus as the parasympathetic physiological system par excellence, related to our altruism and our willingness to take care of others.

Dr. Barbara Fredrickson, a psychologist at the University of Michigan and author of *Love 2.0* (Penguin Group, 2012), has studied how positive emotions (such as love, joy, gratitude) promote creativity and social bonds. She speaks about how "micro-moments" of connection between us expand our awareness while they stimulate the vagus nerve, which in turn slows our heart. She recommends looking for those micro-moments of connection and adding them to a routine of exercise and healthy eating to regulate inflammation, lower sugar levels, and normalize the rhythm of our hearts. It's not optional, it's a biological imperative to connect, Frederickson says.

DiSalvo[85] mentions studies showing that people who regularly activate the polyvagal system by practicing meditation, mindfulness, or deep relaxation tend to be more compassionate and loving, and to

---

[85] D. DiSalvo, "Forget Survival of the Fittest: It is Kindness That Counts," *Scientific American* (September 1, 2009).

experience gratitude more often. The practice of self-compassion is also associated with stress reduction through the suppression of sympathetic activity and stimulation of the vagus nerve.

There are practical applications in the therapeutic space. Say a patient who is on the verge of an emotional collapse has activated his or her fight-or-flight response. A health professional or, in my case, a therapist can use self-soothing techniques, breathing and awareness of the sensations experienced in the body to activate the vagus nerve and calm her down. Then this person can identify which system was activated in response to stressful thoughts or events. This awareness contributes to patients turning their focus to the present, instead of attacking others, fleeing, or competing. I can teach a person how to visualize herself in a safe space to help her regulate her emotions and shift to a new state of consciousness. This will be complemented with creating, or making good use, if already existing, of available support systems to obtain emotional encouragement.

## Clinical vignette

The parents have given an ultimatum to a twenty-two-year-old patient. She has not yet unbound herself from the mother-child attachment. She expects full attention and support from her parents. She's never kept a job for more than three or four months, and thus, she's been relying on her parents' support. In other words, she hasn't cut her emotional umbilical cord. Sharing with other people is difficult for her; she's usually distrustful of the intentions of others.

She is aware of the rationale behind the parental ultimatum: after living with her parents for several months, she's still unemployed, and she spends most of her time on her smartphone or computer when she's not sleeping or watching TV. The terms of the ultimatum: if she doesn't get a job by a specific date, she must leave the house. Their support will end.

That her parents would give her a deadline is devastating news for her. She sees her worst nightmares being fulfilled: running out of money, sleeping on the street, having no resources and no friends.

The parental stipulations lead to an acute emotional crisis, so her parents agree to pay for therapy sessions. A couple of sessions is too brief a time for a therapist to grasp the dimension of the patient's anguish. Because there's a crisis, this isn't the best time to focus on unraveling the root cause of those symptoms that started during adolescence with the use, and sometimes abuse, of substances. She engages easily in power struggles with others, experiences depression and anxiety, and lacks the energy to perform simple tasks. The symptoms explain why she's unemployed and isolated. I see the patient going into an acute crisis mode as the deadline imposed by the parents approaches. She feels stuck at a dead end. She did find a job a couple of weeks ago but she'd need to move to another city. However, without the financial support of her family, she'll not able to pay the security deposit plus first and last month's rent required by most landlords. How is she going to move and take the job if she doesn't have a place to stay? When the job opportunity is gone, she starts considering taking her own life.

At her therapist's office, she decides that it's safer for her to seek emergency care. She chooses a hospital where she could receive proper care. The therapist understands how important it is that, even in the middle of a crisis, the patient take the lead, make her own choices, and feel in charge. After three days of hospitalization and medication, her emotions are regulated, she no longer has suicidal ideation, and she is ready to return to the therapist, who encourages her to take further action in search for solutions.

If all she can experience is fear and anger, she'll stay in fighting mode or avoid reality. If she stays in crisis mode, her brain will keep the primary stress response centers on. First, the amygdala will be activated. Then the adrenals and the locus coeruleus, via the hippocampus, will secrete gushes of noradrenaline and cortisol fostering long-term memory of the event. She will most likely respond to events and people with anger, hostility, and, in some instances, avoidance, which is a form of fleeing. However, as she starts engaging more of her prefrontal cortex, her perspective starts shifting (believing that with support, she will be capable of achieving her goals, for example). By visualizing positive and realistic outcomes, she gets ready for action, activating her second response to stress. The reward will come from solutions, answers. By planning, making phone calls, exploring options, even if there is the risk of bumping against some more closed doors, the dopaminergic circuits (reward system) stay activated. Every small achievement will motivate her even more to action. Dopamine will help stabilize her mood.

We go step by step so she doesn't become overwhelmed.

How much money does she actually have? What's her priority?

She needs to buy medicines, she says, but she's concern she won't have enough money to buy them. A reality test is needed. She calls the pharmacy. The medication is actually cheaper than she thought. She places her order.

We move on to the next phase, exploring her support system. How can she find a place to stay? She remembers her friend mentioning someone was renting a room. She makes a call, and as it turns out, the room is not only available, it's affordable, and she can move right away.

And so, little by little, action is yielding results that counteract the effects of her primary stress response. Instead of panicking and functioning on adrenaline and cortisol, she might be in a state in which dopamine and serotonin are back to normal levels in her body. And who knows, maybe the support she's received in the past couple of days, stimulated the release of oxytocin and endorphins that will calm her down. Her neurotransmitters are back in balance.

The next goal is for her to understand that she can actually rely on others, as she did with the therapist, to support her process and to provide her with, at least, information on available resources. With time, she will likely discover that cooperation and relying on others (not depending on, but getting support from) yields not only practical but emotional and physical benefits. Maybe now that she's found support, she'll be open to the possibility of extending her social network and understanding that when relationships are nurtured, the probability of solving problems grows significantly. The story has a happy ending, but of course, she needs to continue working on processing the traumas that have made her amygdala hyperreactive, developing emotion regulation skills, and changing relationship patterns.

## Cells that allow empathy?

*A cobbler used to work in his yard. In the patio of the neighboring house grew a tall tree from where a monkey observed the cobbler's work. Every day, the monkey would climb to one of the tallest branches to observe and then mimic the shoemaker. When the shoemaker hammered, the monkey hammered. If he cut leather with his knife, the monkey imitated the gesture of cutting leather. Bored of the mimicking, the cobbler decided to get rid of the monkey. He tied a balloon full of red ink around his own neck and then threw a blade at the monkey. The shoemaker brandished his knife and mimicked cutting his throat. The monkey imitated him for a while cutting deeper and deeper into his throat. Then the shoemaker punctured the balloon and spilled the red paint. He doused his hand in the red paint and showed his hand to the monkey. The monkey kept cutting deeper and deeper, touching his neck with his hand, but there was no red stuff to show to the cobbler. Finally, the monkey's hand was stained with a red liquid, but he'd gone too far: he had slaughtered himself!*

The principal of my school, Doña Celia, used to tell us this story of the monkey and the shoemaker in her long edifying lectures, which took place every Friday afternoon in our school's great hall. The point of our principal's story was probably to warn us about the dangers of foolishly imitating other people's behavior or giving in to group pressure. Maybe her story was inspired by a very similar tale by a fifteen-century writer, Bonaventure Des Périers entitled "The Cobbler Blondeau."

This cruel story paradoxically illustrates the process that seems to explain empathy. We do have the tendency to imitate others, especially those we admire, respect, or like.

To better understand our potential for empathic responses, we need to discuss another aspect of our brain's anatomy: the mirror neurons. The capacity to experience empathy seems to depend on the activation of brain regions that are involved in our own experience of pain (anterior insula and anterior and midcingulate cortex). We're able to perceive the emotions others are experiencing, and we resonate with them.

There is scientific evidence that the mirror neuron system is present at birth and that small children learn by imitating those around them. This tendency to imitate others would also explain why we're able to understand how others feel. [86]

Charles Darwin believed that witnessing suffering made us uneasy and that, to lessen our own discomfort, we felt motivated to reduce the pain of the other. Darwin also suggested that extending our compassion to other animals was one of our latest evolutionary acquisitions. He saw "sympathy" as one of the noblest moral virtues that could and would extend to all living beings. Darwin's question as to why seeing someone suffering affects us would be answered more than a century later with the discovery of the mirror neurons (*neuroni specchio*, in Italian).

Neuroscience has reached beyond the explanations available in Darwin's time. We've learned that specific neural circuits in our brains account for our capacity to identify with others and explains why we yawn when others do and why we're touched not only by the suffering of a person close to us but even by the pretended suffering of an actor in a movie or a character in a book.

The mirror neurons are a group of cells on both sides of the brain (in the premotor cortex, the supplementary motor area, the primary somatosensory cortex, and the inferior parietal cortex), discovered in

---

[86] EA. Simpson, L. Murray, A. Paunker, and PF. Ferrari. "The mirror neuron system as revealed through neonatal imitation: presence from birth, predictive power and evidence of plasticity." *Philosophical Transactions of the Royal Society B.* June 5, 2014, https://doi.org/10.1098/rstb.2013.0289

1990 by a team of Italian researchers led by Giacomo Rizzolatti,[87] of the Università degli Studi. They were studying which group of neurons controlled the movements of the hands, and for that purpose they'd placed electrodes in the frontal cortex of macaque monkeys. During the experiments they observed that certain neurons in the pre-motor cortex would get activated while monkeys were offered food. By a stroke of luck, they also discovered that the same group of neurons was activated while a monkey was simply observing a laboratory employee eating food. Further studies with magnetic resonance imaging, transcranial magnetic stimulation and electroencephalograms, showed that this group of cells, which they called mirror neurons, also existed in humans. Rizzolatti later found that these cells reflect emotions as well.

In another of his studies, people were scanned while the odor of rotten eggs was presented to observe what areas of the brain activated. The researchers found later that when participants were exposed to images of faces expressing disgust, the same brain areas activated when exposed to the rotten eggs, fired.

Mirror neurons communicate with the limbic system, which regulates autonomic and endocrine function, particularly in response to emotional stimuli. Further studies in social neuroscience suggest that the same neural networks activated by an emotional experience are involved in empathic responses, including a network comprising the anterior insula and anterior midcingulate cortex. Cognitive control circuits (that process context) would modulate one's empathic responses.

The activation of these mirror neurons in response to the acts and emotions of others seems an attempt by the brain to understand what others are experiencing. It would explain why seeing another person smile makes us smile in return and improves our mood. In 1983, Paul Ekman, who is a pioneer in the study of emotions and their relation to facial expressions, described how expressive facial muscle activity could lead to an emotional experience. The same facial muscles involved in the smile contract in response to another person's smile, even when the gesture is not visible to the eye. It seems that observing someone performing a task or experiencing an emotion can be perceived as one's own experience.

We mentioned earlier that babies perceive and imitate behaviors. We also showed that mirror neurons seem essential for the learning of babies by imitation. The same process would explain an empathic response: I observe someone trying to complete a complicated task or

[87] G. Rizzolatti and L. Craighero. "The mirror-neuron system." *Annual Review. Neuroscience.* 27, 2004, 169–192.

someone in distress, and my mirror neurons get activated. I identify with their difficulty, their suffering (even if they are fictional characters). If I haven't been conditioned with the anathema that what happens to others is none of my business, this empathic response motivates me to try to alleviate the pain of someone going through a hard time.

## Chemistry and neural paths of passional and romantic love

*Romantic love is not only a very strong addiction but a universal craving.*

—Helen Fisher

When I went to medical school in the late sixties, science didn't know much about chemical "messengers." The Chinese had talked for centuries about communication between organs, but it took Western medicine quite a while to corroborate that such communication actually existed. The function of the few neurotransmitters that our teachers discussed (the first one was discovered by Nobel Prize Otto Lowei in 1921) wasn't completely clear yet.

Renowned anthropologist Helen Fisher, author of *The Anatomy of Love* (W.W. Norton, 2017), and *Why We Love: The Nature and Chemistry of Romantic Love* (Henry Colt and Co., 2004) is an expert on human behavior and a researcher on the characteristics of love. She proposes that three different but interrelated neurochemical systems modulate sexual activity and are therefore involved in the reproduction of the species. The first relates to the (instinctive) sexual impulse, the second to romantic (emotional) love, and the third to deep attachment (caring, bonding). Note how these three survival systems parallel the three emotional regulation circuits described above.

The first system, involved in the sexual drive (libido or lust), would explain our search for sexual gratification and the human need to find a mate. The euphoria, optimism, and energy that lovers experience at the beginning of a relationship has been linked to the production of high levels of phenylethylamine (PEA), a substance similar to amphetamine, a stimulant. A release of testosterone is also associated with lust in both men and women.

The second system, activated by romantic love, is associated with a kind of obsession similar to that of an addict. The attention of the person focuses on the object of their love. It is possibly mediated by the production of high levels of dopamine and norepinephrine (and low levels of serotonin), which explains not only the euphoria associated with the activation of this system but also the yearning. From the anthropological, evolutionary point of view, Fisher says, the purpose of this type of love is to conserve mating energy by concentrating on a single partner, until impregnation is achieved. Similar behaviors are seen in all kinds of mammals and also in birds.

The third system is associated with attachment and is modulated by oxytocin and vasopressin. It generates feelings of calmness, emotional union, and safety. From an evolutionary point of view, Fisher said, it would explain the tendency to maintain affiliations for at least enough time to complete parental duties.

In the 1940s, psychiatrists reported a couple of cases that intrigued physicians because of the association observed between spontaneous orgasmic experiences, or sexual hyperarousal, and epilepsy. Later, in the sixties, new cases were reported. Doctors found either tumors or brain damage affecting particular areas of the brain in people suffering from that combination of hyperarousal and seizures. Interestingly, in all cases, not only the seizures but the hyperarousal symptoms faded after the tumors were removed. Research started in uncharted territory. Sexual pleasure, it seemed, was not limited to the genitals or erotic zones and was not a simple reflex or response to a certain type of physical stimulation. A connection had been found between the neural activity in the brain and arousal or climax.

Stephanie Ortigue, of the University of Syracuse, New York, and Francesco Bianchi-Demicheli, from the Psychiatry Center of the University of Ginebra, Switzerland,[88] conducted studies using functional MRI to identify areas of the cerebral cortex associated with orgasm, passionate love, and other types of love (such as maternal or unconditional love). It proved to be much more difficult to study sexual responses by screening with MRIs than by using a single electroencephalogram. In one experiment, researchers used photographs to provoke the subject's sexual arousal. When the subjects experienced arousal, they detected activity in the temporal lobe, as expected, but also in the amygdala (emotion control center) and the hippocampus (associated with memory, especially long-term memory). Brain centers associated with sophisticated forms of thinking and self-awareness of bodily sensations and subjective emotional experience (such as the

[88] S. Ortigue, F Bianchi-Demicheli, N. Patel, C. Frum, and J. W. Lewis, "Neuroimaging of Love: fMRI Meta-Analysis Evidence Toward New Perspectives in Sexual Medicine," *Journal of Sexual Medicine* 7 (11) (Nov. 2010): 3541–52.

anterior insula in the limbic cortex) are also activated. Even areas in the brain associated with empathy appear to be linked to sexual desire. These findings confirmed that the sexual experience is not limited to neural pathways associated with genitals or erogenous zones.

Love circuits are made up of the following brain structures:

1. **The mesolimbic dopaminergic circuit** (or pleasure and reward circuit), of which we already spoke when explaining the systems with which we respond to stress, comprises the ventral tegmental area (VTA), the nucleus accumbens (NA), the nuclei of the raphe, and the globus pallidus in the deepest part of the brain. The circuit is activated when experiencing either love or sexual desire and seems to play an important role in animal attraction.

2. Certain areas of **the prefrontal cortex** related to perception, reasoning, and judgment (executive function) are involved in processing information about partners, the sensation that others complete or complement us, and focusing attention on the love object with exclusion of others. Just as the cerebral cortex correlates with the logical aspect of love circuits, the pleasure and reward circuit can be correlated with some of the emotional aspects of love circuits.

3. **The anterior insula**, located in the conjunction of the frontal, parietal, and temporal lobes has a monitoring function and helps make sense of sensory experience (messages coming from the body).

4. **The hypothalamus and amygdala**, with their important roles regulating emotions and levels of sexual desire, influence how we experience love and the formation of both positive and negative memories.

5. **The thalamus**, as we said, distributes information from the senses to different parts of the cortex to be processed, and plays an important role in selective attention, after processing visual and auditory information.

When people fall in love, they usually build an exalted concept of the object of their feelings, thinking about the other almost obsessively. They can become intense, possessive, and constantly desirous of the other (craving the other). Pleasure, which derives not only from sexual activity but from sharing and from feeling connected and supported, contributes to creating an attachment and maintaining a bond.

Among those researchers studying the circuits of love are Bianca Acevedo and her colleagues at the Albert Einstein School of Medicine, in New York. In one of their studies, ten women and seven men were scanned with functional magnetic resonance imaging (fMRI). They found the VTA was activated when the subjects were shown images of their loved ones. As already mentioned, the VTA is associated with the experience of pleasure and reward, and with dopamine production. Researchers also identified that the nucleus accumbens became activated when people separated from their loved ones.

Donatella Marazziti et al. (University of Pisa in Italy) studied twenty-four men and women who had recently fallen in love and found that the levels of testosterone had decreased in men but had gone up in women.[89] Cortisol was elevated in both men and women, indicating that courtship is stressful. Interestingly, other studies also found that testosterone went up in men after they'd ended a relationship.

Dopamine and adrenaline are also secreted in the brain as love progresses from initial infatuation to a stage of romantic love and to a consolidation of the bond (which is characterized by a dedicated and exclusive relationship). Dopamine is associated with the feelings of euphoria typical of the encounters with or thoughts of the other. Serotonin is found in lower levels when love is just germinating, which might explain why people in this stage of love experience a certain melancholy and lack of appetite.

Oxytocin is also involved in the process of falling in love. As we already mentioned, it's secreted by the hypothalamus and is associated with bond creation. It increases with the physical manifestations of love: kissing, caressing, hugging, orgasm. The elevation of the hormone vasopressin is associated with social behaviors like trust, empathy, and sexual bonding. Curiously, experiments done with promiscuous male voles (a rodent) showed that they became monogamous after researchers found a way to transmit the vasopressin receptor gene, boosting the number of receptors in the reward circuits of the vole's brain.

Pharmaceutical companies are funding studies on these love circuits and genetic variations associated with love behaviors, hoping to find remedies for dysfunctional relationships, love addictions, or to lessen the depression that often appears after breakups.

## Evolutionary advantages of empathy and love

The Polish philosopher Johann Gottfried von Herder coined the term *Eeinfühlung or empathy* in the eighteenth century. He described a method for working across cultural divides, believing that "feeling the subject" (literal translation from German would be "feeling one's way in") would help bridge the gap between the interpreter (of a work of art)

---

[89] D. Marasiti and D. Canale. "Hormonal Changes When Falling in Love" *Psychoneuroendocrinology Journal,* 29(7):931-6 (Aug. 2004)

and what was being interpreted. *The Stanford Encyclopedia of Philosophy* says, "It was...Theodor Lipps (1851–1914) who scrutinized empathy in the most thorough manner. Most importantly, Lipps not only argued for empathy as a concept that is central for the philosophical and psychological analysis of our aesthetic experiences. His work transformed empathy from a concept of philosophical aesthetics into a central category of the philosophy of the social and human sciences."

French neuroscientist Jean Decety[90] points out that in speaking of empathy, we should differentiate between three distinct processes that we usually mix: a) our ability to share emotions, b) empathic concern that leads us to support vulnerable people, and c) the ability to put ourselves in the minds of others and imagine how they feel or what they think.

Nancy Eisenberg,[91] one of the world's leading specialists in empathy and child development and a professor at Arizona State University, defines empathy as "an affective response that stems from the apprehension or comprehension of another's emotional state or condition, and which is identical or very similar to what the other person is feeling or would be expected to feel."

Empathy develops throughout life, if conditions are favorable. Children can often share emotions, but in the stages where their cognitive processes are still egocentric, they have trouble putting themselves in the shoes of another or feeling empathic concern. (We'll later talk about Martin Hoffman's stages of empathy development.)

Psychology usually establishes the existence of at least two types of empathy: cognitive (the ability to intellectually understand what the other person is feeling) and affective or emotional (the ability to feel and respond to what the other person feels, also known as emotional contagion). Those two types of empathy can be mapped neurologically. Emotional empathy is compromised in people with lesions of the lower frontal gyrus of the brain (in the somatosensory association cortex of the parietal hemisphere), and the cognitive form of empathy is affected by lesions of the ventromedial prefrontal cortex.[92]

The capacity for emotional empathy determines the ability to be altruistic. Empathic concern, when observed in a young child, predicts

---

[90] J. Decety, T. Wheatly, eds., *Moral Brain: A Multidisciplinary Perspective* (Cambridge, MA: MIT Press, 2017).

[91] Nancy Eisinger, "Empathy-Related Responding: Links with Self-Regulation, Moral Judgment, and Moral Behavior," Online at https://bit.ly/2P8Td9T.

[92] Paul Ekman, *Moving Toward Global Compassion* (San Francisco: Ekman Group, 2014).

prosocial behavior. Effective, compassionate parenting and education can determine our overall ability to empathize.

A report by a group of scientists from England, France, and the United States suggests that the degree of empathy we're able to experience depends at least partially on our genes.[93] About forty-six thousand volunteers participated in the *Genome-wide analyses of self-reported empathy*, which also found women tended to be more empathic, but that the difference could be due to factors other than genetic—probably hormonal or cultural.

Sympathy, often used as a synonym for empathy, has been described as a response to the suffering of others, but it's actually an unwanted feeling and is based on commiseration. Empathy, on the other hand, is more of an emotional response that recognizes and tries to understand the suffering of the other through emotional resonance ("I feel your pain"). Compassion highlights the most prominent aspects of empathy, adding an altruistic inclination; it's not limited to feeling but involves acts of kindness intended to alleviate the other person's suffering. Paul Ekman talks about four different kinds of compassion: empathic compassion, action compassion, concerned compassion, and aspirational compassion, which differ according to the kind of suffering one intends to alleviate. Based on the target of our compassion, Ekman also distinguishes between familial, stranger, global, heroic, and sentient compassion.

As difficult as it is to define, I will say love is what we feel and express in a relationship in which empathy is experienced and compassion is practiced. It's related to a very complex neurobiological phenomenon that includes not only cognitive aspects that are processed in the cerebral cortex, but the activity of the above-mentioned circuits of the limbic system and cortex that participate in our experience of pleasure and our search for rewards. When we talk about love, we're referring to a feeling that has a certain constancy in time, that motivates us to take care of and protect others because we feel bound or attached to them.

Compassion enters through the door opened by empathy. Compassion goes a step further. Empathy is what allows me to experience and understand how someone else feels and to imagine how would I feel in their situation. If we can feel the discomfort of another person, this feeling can motivate us to support and help. Compassion is feeling with another person (or being), sharing their feelings,

---

[93] V. Warrier et al., "Genome-Wide Analyses of Self-Reported Empathy: Correlations with Autism, Schizophrenia, and Anorexia Nervosa," *Translational Psychiatry* Vol. 8, No. 35 (2018).

accompanying them and trying to alleviate their suffering, expecting nothing in return. Empathy and compassion are required in order to build an ethical, democratic, and supportive society.

In this book, we're using the terms empathy, compassion, love, and solidarity as a group of interrelated but distinct emotions and feelings.

According to psychologist Martin Hoffman, of the University of Michigan, the ability to experience empathy is at the root of our morality, but does it represent an adaptive advantage? Apparently, it does.

Darwin believed that sympathy was our strongest passion. Ekman explains that Darwin used sympathy as equivalent to the terms empathy, altruism, or compassion.

Some Darwin followers claim that compassion and cooperation have been far more important than competition in the process of ensuring the survival of *Homo sapiens*. Not many people know that it was not Darwin but his follower Herbert Spencer who coined the term "survival of the fittest" after reading about Darwin's natural selection theory. It seems that Spencer and a group of social Darwinists used Darwin's ideas to sustain the superiority of a particular class and race over others.[94]

Psychologist and Professor Emeritus at the University of Tennessee Daniel Batson has worked on a well-known empathy-altruism (1991) hypothesis: we are inclined to help others even if the cost is greater than our personal gain. That is, the prosocial inclination motivated by empathy felt toward a person in need has as the ultimate purpose benefiting the other and not the altruist. Other hypotheses postulate we might be motivated to be altruistic because of the joy and pride experienced by helping others.

Our capacity for empathy and compassion may have developed as a response to the challenges in the environment. Goetz, Keltner, and Simon-Thomas[95] hypothesized that compassion evolved as an emotional experience that had the primary function of facilitating cooperation and the protection of the weakest and the suffering. The authors believe compassion affects the process by which people's moral judgments and actions take shape. They point out how this process varies in different cultures and between groups.

Genetics, neurobiology, and evolutionary psychology seem to support this idea. The same authors cite Darwin's description of the tenderness and compassion displayed by the natives of Patagonia

---

[94] See Sylvia McLain's article "Evolutionary Theory Gone Wrong," The *Guardian* (May 2013).

[95] J. L. Goetz, D. Keltner, and E. Simon-Thomas, "Compassion: An Evolutionary Analysis and Empirical Review," *Psychological Bulletin* 136(3) (2010): 351–374.

toward the members of their families. He theorized that, at the beginning, we might only have felt compassion for those closest to us, observing that those same natives who showed compassion toward their offspring allowed their children to learn how to torture prisoners of war. Over time, humans would have learned to feel compassion toward clan members, extending their empathy to those of the same tribe, and later developing the capacity to feel compassion for their people, ethnic group, co-partisans, and finally the whole of humanity.

In his controversial book *The Selfish Gene* (Oxford University Press, 1976), author Richard Dawkins stated that it's the genes striving for immortality that drives the evolutionary process, and individuals, families, and species are merely vehicles in that quest. For example, the worker bee does not pass on its genes directly, but its work contributes to the well-being of the queen, which guarantees that the genes they both share could be passed on to the next generation.

The paradox, scholars say, is that the selfish gene would motivate us to be altruistic—to save others—but just because we share their genes. According to this theory, species change because individuals feel the urge to thrive at the expense of others.

In *The Age of Empathy: Nature Lessons for a Kinder Society* (Three Rivers Press, 2009), Frans de Waal, from Emory University, disagrees with Dawkins. He believes that we have an innate sensitivity that allows us to capture the needs and feelings of others, even if such sensitivity is not always evident and although acting on it is not always in our best interest. De Waal, who considers natural selection from the perspective of the group, argues that empathy is crucial because it serves as an adhesive to societies and guarantees that the sick, the children, and the elderly will be cared for. In addition, in urban life, says de Waal, where we need to learn to tolerate thousands of strangers, empathy greatly contributes to coexistence.

In his book, de Waal tells the story of the August day in 2005 when he was driving from Atlanta to Alabama. It was the day following Hurricane Katrina's US landfall, when the floodgates protecting New Orleans failed, and 80 percent of the city was flooded. Thousands of people became homeless overnight, and de Waal witnessed the flow of people to Alabama, where they overcrowded the hotels. The levees had yielded not only because of the force of the hurricane and the torrential rains but because the levees had not been properly kept up. The maintenance monies had apparently been diverted. Katrina has been one of the costliest disasters in the United States (estimated 161 billion dollars), and although the trigger was a natural phenomenon, the real cause was both an engineering failure and administrative corruption.

In the aftermath, we also witnessed the memorable performance of Louisiana's local emergency personnel; many risked their lives to save others. Ordinary citizens performed notable acts of courage. People lent their yachts, boats, and canoes to serve as emergency ambulances and to rescue those trapped in flooded areas. Many concerns remained in the air, the most pressing, in addition to the lack of prompt response from president Bush and the federal government, was how the maintenance of the levees was neglected without considering the common good? How could the greed and neglect of some destroy the lives of others? What happened to their empathy? How is it that the calamitous consequences of this disaster in the lives of thousands of people were not anticipated and prevented?

These reflections motivate me to share another key question that de Waal raises in his book: Could a thriving economy and a more humane society coexist? I would also ask if the drive to amass money doesn't destroy empathy. I think the answers would depend on how we define prosperity and progress and what we mean by a more human society.

A professor of Psychology at the University of California at Berkeley, Paul Piff, studied the relationship between material wealth and narcissism. He observed participants in a rigged game of Monopoly, in which randomly chosen players were given certain advantages (more money, more opportunities to roll the dice). During the game, the players with the special advantages showed more narcissistic attitudes, displays of power, and greed.

In an interview for the *Guardian*, a reporter asked de Waal, "So, is it harmful to focus on our alleged selfishness?"

De Waal replied that it was "extremely dangerous." He went on:

> Many economists are great believers in the idea that everything in nature is competitive and that we should set up a society which is competitive to reflect that. Anyone who cannot keep up, well, too bad.
>
> I believe that is a total misinterpretation of the facts. The individual is not all-important. Yes, we can be selfish, but we are also highly empathetic and supportive. These features define us and should be built into society.[96]

If we're to be a more humane society, we have to be sufficiently aware of the repercussions that individual and group acts have on the community; therefore, we need a responsible society that cares for others and especially for the planet on which we depend for our survival. It

[96] R. McKie, "My Bright Idea: Humans Found a Nicer Way to Evolve," *The Guardian* (September 18, 2010).

needs to be a fundamentally empathic, compassionate, and supportive society in which individuals use their abilities to understand the other's point of view, are able to be moved by the suffering of others and are ready to reach out to help them when they're going through adversity.

## Moving toward compassion

The development of empathy seems a sine qua non for being compassionate or developing the ability to love. Without empathy it's difficult to cooperate in efficient ways or to be supportive of each other.

From the evolutionary point of view, the purpose of compassion is to lessen suffering, which is a normal part of life. When we're distressed, we calm down in the presence of another who keeps calm. The serenity of the other reassures us, and therefore we seek out a compassionate person when we're afflicted.

Robert Wright is a US reporter and the renowned author of books about science, evolutionary psychology, history, religion, and game theory, including, *Nonzero: The Logic of Human Destiny* (Pantheon, 1999) and *The Evolution of God* (Little, Brown, and company, 2009). He teaches at Princeton University about the connections between cognitive science and Buddhism. Wright talks about the natural history of compassion. If we've become more compassionate, he argues, it isn't because of the influence of a religion, even though all of the world's major religions view compassion as a fundamental virtue. Before *Homo sapiens*, Wright says, compassion already existed, because compassion is embedded in our human nature, and besides, it isn't exclusive to humans.

Animals also have an affective life: they grieve after the death of a companion or an offspring, and, in the case of domesticated animals, they grieve when their master leaves or dies, and they even give signs of identifying with the state of mind of those who care for them. Certain geese show grief when they lose their partners. Sea lions sadly moan when they see their children perish in the face of a whale attack. Dolphins try to resuscitate a dead infant and then bellow.

Human consciousness is not as unique as we assume, although animals do not articulate their feelings through spoken language. We surely can learn something from the way many animals feel, offer each other support, and learn to cooperate.

Famous biologists who advanced evolutionary theories in the early last century, such as J. B. C. Haldane and Sir John Maynard Smith, observed how compassion was expressed between animals. A compassionate animal is inclined to help others (generally when recognized as part of its group or family), which may offer a long-term benefit to the genetic group. Another related behavior apparent among animals is the so-called reciprocal altruism, in which an organism risks its welfare to improve the well-being of another, apparently with the goal of receiving reciprocal treatment in the future.

In their review on the evolution of compassion,[97] Goetz and colleagues integrate three arguments that support the hypothesis that compassion evolved as a distinct affective function to facilitate cooperation between members of a group and protect its more vulnerable members. They define compassion as the feeling that comes from witnessing the suffering of another and that motivates a subsequent desire to help. By defining it as a feeling, they distinguish it from other definitions that regard it as an attitude or behavior, and they also differentiate it from empathy, which would refer solely to the vicarious experience of the emotions of another being, what we commonly call putting oneself in someone else's shoes. The authors do recognize sympathy, empathy, and pity as emotions related to compassion.

So how does compassion arise as a component of love? Theories differ. Drawing form different sources,[98] I offer a summary of three of them:

1. Compassion contributes to the well-being of the most vulnerable. Evolutionarily, it's believed that the group and the individual do whatever is necessary for the preservation of their offspring. This would explain maternal love.

2. It's a desirable emotion when it comes to mating. Hopefully a compassionate person will be a better parent and take better care of the partner. Sensitivity to the needs of another, sensitivity resulting from the capacity to feel compassion, contributes to more stable affective attachments. This would explain love between couples.

3. It improves the chances of cooperation with others, from the same group or external groups. The society rewards altruism with integration and punish the selfish with social isolation. This would explain unconditional love.

---

[97] J. Goetz, D. Keltner, and E. Simon-Thomas, "Compassion: An Evolutionary Analysis and Empirical Review," *Psychology Bulletin* (May 2010).

[98] T. Esch and G. B. Stefano, "The Neurobiological Link between Compassion and Love," *Medical Science Monitor*17(3) (2011): 65–75.

Theories about the evolution of love presuppose that human emotions have emerged as the result of an adaptive process in response to situations related to the survival of the species.

Love (understood as a feeling that comprises empathy, compassion, and solidarity) has given humans an evolutionary advantage; therefore, the ability to love has been transmitted from generation to generation. An opinion shared by evolutionary psychologists is that the golden rule (treating others as you would have them treat you) is intuitive.

The forms of love most clearly characterized by compassion are maternal love and romantic love, but the expression of compassion is not intended to create attachments and doesn't depend on the attributes that attract us to another person. On the contrary, compassion comes from that empathic response that allows us to respond to suffering without expecting reciprocity.

## Unveiling the mystery of love

*The most beautiful thing we can experience is the mysterious. It is the source of all true art and science. He to whom the emotion is a stranger, who can no longer pause to wonder and stand wrapped in awe, is as good as dead—his eyes are closed.*

—Albert Einstein

We've seen love become a subject of study in the world of science. One of the main questions motivating researchers is, what purpose does love have? Anthropologists, geneticists, and evolutionists argue that love ultimately guarantees the reproduction of the species. Those who study attachments believe that our ways of loving and relating are the product of early childhood experiences. And neuroscientists explore what happens in our nervous system when we experience empathy, compassion, and love.

The truth is we know very little about the natural development of compassion and love in children and adolescents or about how parenting styles contribute to the socialization of children or the development of empathy. We don't even have clear, unified definitions of what love and compassion are, and such definitions are needed to devise unified parameters for measurements. However, we're certainly moving forward. Let's mention some examples of systematic work in the social sciences that have contributed to a better understanding of the nature of relationships and the process of learning empathy and love.

Arthur Aron and his wife, Elaine, have been pioneers in the study of passionate love for the past fifty years. Aron is a professor of psychology at New York University. The couple became famous for thirty-six questions they came up with to be used in their laboratory to help couples develop a greater degree of intimacy. But not just couples. The questions are so effective that they even amuse total strangers who sometimes start relationships that end in romance after answering the quiz. Self-disclosure, the researchers believe, grants the success of a relationship.

Dr. Gary Chapman, an expert in marriage, has theorized that there exist five kinds of behaviors (or languages) with which couples express love. These languages, according to Chapman, are: a) using words of affirmation, b) sharing quality time, c) giving gifts, d) acts of service, and e) touch: expressing affection physically. If our languages are different, we'll have trouble understanding each other. He's designed a questionnaire to help people be conscious of the ways they and their partners understand love.[99] I propose that these same behaviors are not exclusive to romantic or passionate love but also characterize other forms of personal love: toward our children, friends, and parents.

Goff, Goddard, Pointer, and Jackson[100] determined the validity of the parameters used by Chapman, concluding that they could effectively predict the marital satisfaction experienced by the study subjects. Chapman's idea is that even if we have different love languages we can still get along fine. However, understanding our loved one's needs and expressions and, even better, learning to speak their same love language, could greatly improve our relationships.

Over the course of almost fifty years, hundreds of couples have participated in Dr. John Gottman's studies. In his love lab, in Illinois, equipped with video cameras, blood pressure and pulse sensors, and computers, Dr. Gottman and his colleagues have studied the conversations couples hold, their interaction patterns, their tone of voice, their body expression, as well as micro-physical and emotional expressions that provide key data regarding what holds or breaks their bonds.

Researchers, including Paul Ekman, have suggested that if people are experiencing any kind of love, it can be inferred by studying their emotions, behavior patterns, cognition, facial expressions, and physiology.

We already mentioned Nancy Eisenberg, a pioneer in the study of the development of empathy in children. For forty-five-plus years, she has contributed to the understanding of our biological makeup

---

[99] G. Chapman, *The Five Love Languages*: *The Secret to Love That Lasts*. Chicago, IL: Northfield Publishing.

[100] Ibidem.

(temperament) and how we regulate emotions. She's tried to apply her findings to the way we parent children to promote pro-social behavior.

Clara Strauss and her colleagues[101] reviewed the scientific literature trying to find psychometrically valid measures by which to evaluate compassion. By consolidating existing definitions, they proposed that five elements characterize compassion: a) the capacity to recognize suffering, b) understanding the universality of human suffering, c) feeling the suffering of the other, d) tolerating feelings that make us uncomfortable, and e) motivation to act in order to alleviate the suffering of another.

The drawback is that science needs quantifiable data to prove the accuracy of researchers' assertions. Without unified definitions and appropriate measurement tools, it's difficult for scholars to study partner interactions, measure feelings, or assess whether the interventions used to nurture these emotions are effective. Social science has always struggled to measure things that are as subjective as emotions and relationships.

I am, however, confident that more screening tools for empathy—similar to those already in use to assess autism in adults or children's developmental milestones in the motor, cognitive, psychosocial, and communication domains—will soon be available everywhere. More than being focused on behaviors and mental health, these will be instruments used to track the development of empathy, compassion, and even something as subjective as the capacity to love self and others. We will be able to measure, for example, our inclination to please another person, tune into their pain, or attend to their needs, the time we want to share with the other, our degree of interest in their contingencies, all criteria with which, if not yet by science, by common sense we commonly identify with feeling love.

[101] C. Strauss et al., "What Is Compassion and How Can We Measure It? A Review of Definitions and Measures," *Clinical Psychology Review* 47 (July 2016, ePub May 26, 2016):15–27.

## PART FOUR: Our Planet, a Loving Place?

*Love and compassion are necessities, not luxuries. Without them, humanity cannot survive.*

—Dalai Lama

*When they left home, their plan was to drive to a nearby beach and take a quick walk and a swim. Roberta Ursrey, her mother, her nephews, and their children, enjoyed going to the beach, walking barefoot on the sand, feeling the breeze near Miller Pier in Panama City Beach, Florida. When they arrived at the beach, they all noticed the yellow flag, warning of the presence of strong rip currents. Roberta warned the kids to be careful, and then ran barefoot across the boiling sand and took a quick dip. When she got out of the water a few minutes later, she realized the children were in the water but had moved away from the beach, farther than advisable. When she heard their screaming, she realized that they were being dragged by a rip current. Roberta and other members of her family also wanted to swim toward them, to rescue them, but the riptide was very strong, and they were also dragged. While being driven by the current, Roberta's mother suffered what seemed a heart attack. At that moment, the panic of losing her whole family overwhelmed Roberta.*

*While this was happening, Jessica and Derek Simmons noticed that many on the beach were looking in a certain direction. There were police patrol officers, and Jessica thought: Shark! She touched her husband's shoulder and nodded; they ran into the crowd and then saw that about a hundred feet away, some people seemed to be drowning. Jessica was a woman of action. She'd once walked nearly twenty miles in Alabama after a tornado to aid affected people. She was an excellent swimmer, and when she realized that nine people had been caught in the current, she grabbed her boogie board and with no hesitation threw herself into the water. On shore, people started to grab hands, linking up wrists, legs, arms, forming a human chain, that reached about eighty people, with the intention of pulling the swimmers back to safety. After an hour, all those washed away by the current had been rescued. Roberta's mom was transported to a nearby hospital by an ambulance. (Different sources)*

I wanted to start this section with a story of solidarity taken from real life. Many more stories like this are often overlooked. Unfortunately, the emphasis of the media feeds our negativity bias, so we tend to view the world as mostly insensitive, selfish, and violent.

However, in the midst of this turbulent world in which we live, thousands of displays of empathy, compassion, courage, and remarkable solidarity in the face of tragedy are constantly happening. Often, they involve sacrifice.

It's January 2019 as I write this sentence. Still fresh in my mind are some of the worst world news stories that have touched us in the past two years. The slaughter at the Orlando Pulse club (fifty dead). The massacre by Stephen Paddock of people attending a concert in Vegas, (fifty-nine dead, five hundred wounded). The bomb at the concert of Ariana Grande in Manchester, England (twenty-three dead), and the attack on three policemen on the Champs Elysees in Paris (three dead). In November 2017, Egypt suffered one of the bloodiest attacks on a mosque (305 people killed). There was also the massacre of seventeen students at a school in Parkland, Florida, on Valentine's Day, 2018, a tragedy that has mobilized youth around the world against these acts of violence and for gun control. In November 2018, the Tree of Life synagogue in Pittsburgh was assaulted, and eleven people were gunned down. But also, even as I write, thousands of other tragedies occur in unknown villages in countries like Colombia, Venezuela, Yemen, and Palestine. And anonymous people respond with acts of courage and sacrifice to this senseless violence. We seldom read about it in the news.

Contrary to many studies showing empathy is declining and egocentrism becoming more prevalent in the world, Steven Pinker argues that cosmopolitanism has led to a more empathic human being. Perhaps the movements that germinated in 2016 and 2017 in the United States[102] protesting for justice, for civil rights, and for better wages, or vigils such as those seen in Colombia in 2018 in protest for the murder of hundreds of social leaders, are an indication that our capacity for empathy is flourishing in the midst of all this social turmoil in which we are immersed.

We've also seen displays of solidarity and empathy in response to recent natural disasters. After Hurricane Irma hit Southwest Florida in September 2017, with all the force of its winds and rain, examples of

---

[102] In the article "Rallying Nation" (April 2018), the Washington Post reported that since the beginning of 2016, one in five people took part in political rallies in the United States, of which 20 percent had never participated in politics before.

solidarity multiplied. On Facebook, many people shared stories of how, in the midst of the muddle, people opened their houses to lodge and feed strangers. Generous donations and in-kind contributions were made to help those who were hit the hardest. People used social media to share information about, for example, available shelters, hot meals, electricity, or the locations of gas stations that still had fuel. Those who had evacuated before the hurricane had their neighbors or friends checking up on the damage their homes had suffered.

Almost a month after Irma, another hurricane, Maria, devastated Puerto Rico, in late September 2017. Rosana Guernica, twenty-two, a Carnegie Mellon student, made the decision to act on behalf of her compatriots who were incommunicado, without electricity or water, and without access to essential services. Using a crowdfunding site on the internet, she made an urgent appeal to support what everyone thought was a crazy idea: raising enough funds to rent an airplane and bring aid to the island. Despite the disbelief of many, she raised more than enough funds. The private plane she hired not only carried medical supplies, food, and water to the island, it also brought people with urgent health needs back to the continent. She didn't do this alone; a group of enthusiastic volunteers supported her.

The world evolves because of the contradictions of sectarianism-violence and union-solidarity. There is dynamic interaction between opposites. Transformation and metamorphosis are brought about not only by the contradictions between opposing forces or aspects of nature or productive forces, but also by the collaborative interaction between these aspects (symbiosis, parasitism, mutualism, commensalism).

The Chinese Taoists explain the concept of Tao—the all—that gives rise to polarities, which they call yin and yang. The concept is symbolized by a circle containing two opposite sections: white, which moves and grows to becomes its opposite—black. And black, which grows toward white and transforms into its opposite. Each aspect always contains the seed of the other. This concept represents the duality of everything that exists in the universe. The yang would represent a luminous energy, positive, active, and the yin a negative, passive energy. This dialectical concept helps us understand the complexity of a universe that's always in motion.

When I look around, both in my professional life and in my social and family life, I see memorable acts of courage and solidarity. Despite the wars, famines, generalized violence, and displaced people, there are also admirable people, like the protagonists of the brief stories I've shared, people performing acts of love at this very moment. They are there, even if you don't see them in the news or on the social networks.

It would be impossible to show all the good that humans are able to do and are doing. It's just not very newsworthy.

In his book *The Empathic Civilization: The Race to Global Consciousness in a World in Crisis (TarcherPerigree, 2009)*, economic theorist Jeremy Rifkin points out that although man has a gentle nature and generally lives in peace, history and the news continually transmit an image of a violent and conflictive society, and our collective memory ends up being measured in terms of crisis and calamities. Analyzing the reasons for this phenomenon, Rifkin proposes that the stories of the infamous or illicit acts surprise us precisely because they break the norm. For that reason, they're better news material, and they feed history. No matter how much turbulence we've known about in recent years, I cling to my faith in humanity and believe this is true.

## The awakening of love

That which inspires, obsesses or distracts us, that which motivates us to move forward, that which feeds our aspirations, that which gives us the blues or helps us believe in ourselves, and ultimately, all things in our lives seem to be related to some aspect of love. Most of our suffering comes from the lack of love, from our failure to fully engage with life, to find our purpose and dynamically belong to the world. It's our inability to connect with our bodies, with our own emotional needs, and with other people that affects our physical, emotional, and mental health.

Many of the people who seek my counsel have experienced rejection, abandonment, or betrayal, all leading to some form of heartbreak. Being deprived of love has marked their lives forever, making them afraid of engaging in a new relationship. It's caused them emotional pain and determined their (sometimes limiting) relationship patterns. So, they seek help to alleviate their suffering and process their traumatic experiences. Some need to learn self-compassion. Then they're able to step out of their shells and lean on others for support. Others find new joy through service. A few, finally, forgive or reconcile with a loved one. They eventually heal their minds and their lives. The fact that they seek a counselor's help doesn't make them a separate group. On the contrary, heartbreaking experiences are universal. Suffering has affected everyone at one time or another. And part of that pain comes from not having learned to feel compassion for oneself and others.

Facebook, Instagram, Snapchat, and Twitter may be contributing to changing our perspectives on life and the world. Behind those lovely photos and moving videos shared to show us uncommon, extraordinary, or cute aspects of other humans or of animals, we can perceive someone's awe. What's most interesting in the posts is the person who shares them: someone with enough sensitivity to observe, admire, value, love, and even long for experiences of connection, friendship, mutual care, devotion, or concern, found not only among us humans but among animals as well. Social media has also played an interesting role in calling up people to take community action or to protest against injustice, advocate for social causes, and share key information, which the media often discards, overlooks, or distorts. Unfortunately, sometimes, social media have also been an instrument to spread hate.

However, maybe the internet, with all the risks it poses to a deeper interpersonal communication and introspection, is ironically making us aware not only of the existence of the diversity in humanity but of the million things we have in common with all human beings. Furthermore:

It proves our great need to connect with others.

It shows our need to be seen and validated.

It invites us to be inclusive.

It suggests that being different does not have to be socially risky.

It shows us that respecting animals and nature is not only possible but desirable, if we want to preserve the habitats and species that our modern life has endangered.

It demonstrates that we're interconnected in more ways than we can sometimes see or understand.

This digital era constitutes a mass of free-flowing experiences, information, and interactions—an iceberg so to speak, of which only a small part is visible. The problem is, if this iceberg dangerously tilts against the tenets on which our societies are built, it threatens to transform our lifestyles, beliefs, and relationships. In the long run, it will cause a few unavoidable ideological and perceptual wreckages. It depends on which side this hulk tilts to. It would either alienate us or homogenize us. (Think *Brave New World*, by Aldous Huxley, in which, through eugenics, everybody achieves an insipid happiness.) It could turn our existences dull, uninteresting, and dangerously uniform. Or it might make us shallow, pleasure driven, and narcissistic. But it could just as well contribute to waking us up, encouraging us to take more responsibility, and inciting us to pursue common interests collectively.

So far, it feels as if we're learning to relate in the wrong ways. Not from love but from the selfishness of a misunderstood freedom. Not with empathy but with the expectation of a Hollywood movie kind of love. Perhaps, as disturbing as it could be, we're more and more taking refuge in loving our domestic animals. At the end of the day, pets don't reject, threaten, betray, or abandon us. Our love for our pets symbolizes, I believe, the search for a rather undemanding, comfortable love. We like the docility of the animals we adopt, their gratitude and loyalty. We need them and the coziness they offer. We can expect to be respected, obeyed, and followed. Besides being cute, pets move us to tenderness. They make our days happy and allow us to express love safely. They become an "object that does not object" to the treatment we give them. A pet is a submissive animal by definition. It will not discuss our decisions or criticize or judge our behaviors. No nagging coming from them. We believe they love us back, even though their apparent resonance with our mood is likely due to what researchers call "mood contagion," a reflex reaction similar to what causes a flock of birds at the beach fly in unison as soon as someone comes by. Instead, loving another human being involves a much greater challenge that requires work that includes practice, acceptance, empathy, respect, and definition of clear limits.

Let's not forget that behind every cute and faithful animal that moves our heart and inspires our awe lives a human capable of offering their love, care, and attention. A human whose heart can feel tenderness toward another living being. A person who's probably rescued, fed, and sheltered those pets affectionately. One who triggers the gratitude expressed by them. And in the virtual world, behind each *like* and each *share*, there's someone who can be tender and whose heart is moved by the cuteness of pets and their demonstrations of affection.

Another interesting dimension of this new relationship we have with animals is that vegetarianism is increasing (from about 1 percent to almost 6 percent in the US in the past two decades), and we've also become increasingly aware of the importance of consuming products that don't involve any form of animal abuse. Because we've learned about factory farms where animals are often kept in overcrowded, filthy, windowless barns, or stuffed into cages and metal crates, more and more people are avoiding eating eggs from hens that have been corralled the whole day, or meat and poultry from cows and chickens that have been treated with hormones or steroids. When using supplements, medicines, and cosmetics, we now make sure the products have not been developed using research on animals. And this is not just because of personal health reasons but because we're increasingly conscious that the former anthropocentric ways in which humans have been relating to other living beings and the environment is wrong.

For a bodhisattva (in Buddhism, a saint who chooses to continue incarnating indefinitely in order to serve humanity), caring for all living beings is a fundamental principle. Ekman reminds us that this compassionate concern for all sentient beings is exclusive to Buddhism and does not appear in Abrahamic religions (Judaism, Christianism, and

Islam). In his most recent book, *Moving toward Global Compassion* (Paul Ekman Books, 2014), Ekman asks why everyone isn't concerned about the welfare of all people, everywhere.

What seems hopeful is the certainty that those human qualities we've just mentioned aren't limited to relationships with the animals or to activity on the internet. We have the capacity to love and care for others, to show solidarity. Even the sharing of a video, a touching photo, or an inspiring quote with our friends seems to be an attempt to give, connect, enjoy, share, or (symbolically) be on the same page with them.

As we've seen, neuroscience supports and is validating the notion that we were born equipped for love and compassion, that our physiology is encoded to feel these emotions, and not only toward our partners, our children, or our parents. We human beings have a basic need to feel that we belong to a group, that we're connected. In a globalized world, compassion should also be global, as Ekman puts it.

Given the state of affairs in the world—when civilization is not defined by the progress in our relationships with other humans, with other living beings, or with the planet—it's difficult to believe we have a loving essence.

We've now established that there is sufficient evidence from a biological, mental, and even spiritual point of view to assert that human beings are endowed with the genetic, anatomical, and physiological components to experience love and empathy. Feelings of compassion incline us to solidarity, to feel the pain of another, to unite, protect, and share, sometimes even putting our own safety and well-being at stake. That's why it's fair to think that love could save the planet.

## The challenge of embracing love and freedom

*I may now add that civilization is a process in the service of Eros, whose purpose is to combine single human individuals, and after that families, then races, peoples and nations, into one great unity, the unity of mankind.*

—Sigmund Freud

This certainty that we're capable of solidarity, compassion, and love for others is the best hope we have in approaching the enormous challenges humanity has ahead of it, which will require great changes and enormous sacrifices. It's our responsibility to participate and contribute to this discussion about how to build a more conscious society and a sustainable economy that's not driven by greed. A society that would provide optimal conditions for us to develop our innate ability to love, be mutually supportive, and cooperate in order to overcome the obstacles we face.

On June 14, 2017, James Hodgkinson, a lonely man who was facing financial difficulties and was disgusted with Republican legislators, opened fire on a baseball team formed by Republican congressmen. They were practicing for the congressional baseball game for charity in a Washington, DC, suburb. In a climate of tense animosity and partisan sectarianism, this would be a friendly match. Hodgkinson arrived on the field, raised his rifle and fired at four people, including Congressman Steve Scalise, who was seriously injured.

After the attack, reactions varied. Scalise himself, after he recovered, pointed out that it was incorrect to politicize what had happened. He called for giving priority to the people who had been injured and their families. Calls for unity were made by both parties.

Bernie Sanders, a presidential candidate at the time, said violence of any kind was unacceptable and change could only come through nonviolent action. His words are not novel. Who should adhere to them? For starters, all those who call themselves Christians, given Christ's cardinal command to love one another.

Mahatma Gandhi, who was raised as a Hindu but who studied many religions, was a great defender of a nonviolent ideology in political practice. He was against all forms of aggression. He would have also spoken against a further polarization and warned us once more: an eye for an eye and the world will end up blind.

The preaching of spiritual leaders such as Buddha, the Dalai Lama, and Pope Francis divulge the same message. Nonviolent action is a powerful tool for social and political change.

There are those who have suggested that a better name for our species would be *Homo poetico*: a hominid that seeks purpose and meaning. However, the species and the planet would benefit largely if we evolved toward a *Homo amandi,* a homind capable of love. After all, as Pope Francis says, love is what gives meaning to life.

The philosophical debate about the nature of man (or an immutable human nature, if there is such a thing) has permeated the academy for decades. Are we the product of our position in society? Are we good by nature and corrupted by society? Is our life governed by the soul, or is it dominated by our biology? Are we naturally connected with all beings? Are we co-responsible for the universe, or are we free only in our individuality? Is our human nature connected with something intangible?

The debate is relevant when we propose that man has evolved to be physiologically equipped for empathy and love and, therefore, for cooperation and solidarity, but that in a world defined by a voracious, competitive, frenetic consumerism, a globalized world led by individualistic goals, it's difficult to cultivate kindness and compassion, and especially love.

"Power is of two kinds. One is obtained by the fear of punishment and the other by acts of love. Power based on love is a thousand times more effective and permanent then the one derived from fear of punishment," Mahatma Gandhi said.

It's ironic that the predominant economic system has led us away from the freedom it preaches. It has alienated man, leading to the loss of our humanness; it has transformed us into objects. When the only capital a person owns is their work, there's a risk of becoming objectified, and the freedom of expressing ourselves through our doing becomes an illusion. The opposite of objectification is self-affirmation, which leads to self-love and real freedom.

The expression of feelings is often seen in Western as a sign of lack of control, emotivity, or a vulnerability—not be exposed in public. The human being has suffered a process of dehumanization and automation, mostly as the result of industrialization, the conflict between world powers, and the relationships between those who own the means of production and those who owe nothing. In his book *Marx's Concept of Man* (Frederick Ungar Publishing, 1961), psychoanalyst Erich Fromm had already predicted that if we continued on the course we were on at the time, the historical perspective of the Western world would be bleak, and we would be facing physical or spiritual extinction.

More recently, the late physicist Stephen Hawking also warned us about threats facing humans and the Earth, including artificial intelligence, climate change, genetically modified viruses, and nuclear war.

How can we overcome such tragic destiny?

## Made for love but...

How could we restore our totality if it's not, as Marx said, by finding unity and harmony with our fellow human beings and with nature? By recognizing ourselves as members of a society and being aware of the social forces that act upon us, and imprison us, we could become truly free.

Many of the most significant changes humanity has gone through have come through violent action, in most cases as a result of the struggle of the people against their oppressors. Violent was the passage from slavery to feudalism, as in the rebellion led by the slave Spartacus during the Third Servile War, a major slave uprising against the Roman Republic more than two thousand years ago. And violent was the Civil War in the United States that, almost twenty centuries after Spartacus, ended with the emancipation of the slaves (1865) and inspired the liberation of blacks in other American colonies. Let us note that despite the declaration that emancipated the slaves, slavery was not completely eliminated in the country until the mid-twentieth century, and still now, the social disadvantages of different ethnic groups are palpable.

In the United States of the nineteenth century, white people and Indians hunted black slaves together. And, in acts of retaliation resulting from the hatred generated by slavery, many slavers died at the hands of slaves who fled the plantations and headed to Mexico, where the country's independence from Spain was immediately followed by the abolition of slavery (1829). Almost half a century later, while the northern states of the United States promoted abolitionism, the owners of the cotton plantations in the South still clung to slavery, breeding the polarization between slave states and free states, which led to a long, bloody civil war that concluded with the unification of the states.

In the end of the eighteenth century, the French Revolution was an uprising that led to the final liberation of the serfs of the glebe from subjugation by the abusive feudal lords. The revolution violently overthrew the absolutist monarchy, bringing into power a new social class—the bourgeoisie—which brandished the ideals of freedom, equality, and fraternity. From the Bastille in the old continent, the so-called fundamental rights of man were proclaimed and disseminated, serving as ideological food for the revolutionary wars for independence in the Americas.

The English colonies in the Americas were the first to revolt in an effort to shake off the monarchical yoke of Great Britain, declaring independence in 1776. The cost in lives of a revolution that happened before the existence of machine guns or weapons of mass destruction was quite significant.

It's worth noting that the United States has been considered the oldest modern democracy in the world and the first one to use suffrage to elect their representatives. The English had instituted a parliament in 1688, but they never eliminated their monarchy. The Pilgrims, a group of reformist English settlers with strict religious beliefs, arrived in the North of the United States in the seventeenth century and settled in the Plymouth area. They had already created a tradition of governing by nonviolent means, through the establishment of covenants.

In the first half of the twentieth century, Mao Tse-Tung (Mao Zedong) led a war (1937–1945) that shook the whole territory of China and ended the unfair treaties by which foreign parties had felt entitled to station their military forces in both land and sea. Those treaties had led to England and France usurping land within China to exploit natural resources. In 1931, Japan had colonized a big portion of China. When in 1949 Mao took power, he led the creation of the People's Republic opening the doors first to industrialization and later to socialism in what was then a colonial and semi-feudal country. Today, China has become a world power that seems to have embraced many of the ways of Western culture and is apparently taking giant steps to expand its economic interests, invading the global market with their products, establishing alternative energy businesses in Africa, petroleum exploration in South American countries like Venezuela, Ecuador, and Argentina, and developing mining operations in Ecuador and Peru.

Part of the current armed conflicts on the planet originated in what is known as the Arab Spring, a series of riots and protests that occurred in the Middle East in 2011, in which the people, fed up with repression, corruption, and abuse of power by its leaders, rose up, demanding civil rights and democratic participation. Five dictatorial governments fell, including that of the most charismatic leader in the region, Libya's Muammar Gaddafi. Since then, the internal crises of these countries have only worsened, and violence has not stopped. The fire seems to be fueled by private corporations and governments with political and economic interests in the region.

Contradictory as it may seem, the ultimate motivation of insurrections (against oppression and domination) has been, and still is, to open the way to more solidary societies and to social cooperation, which appear to be the evolutionary human destiny.

It's necessary to understand that violence has not been the only means used by the masses in pursuit of their rights or when they have sought liberation from an oppressor. Civil resistance movements and nonviolent action (especially in the last hundred years) have proven to be even more successful than violent wars at creating change. Researchers and professors at the University of Denver, Maria Stephan and Erica Chenoweth[103] found that "major nonviolent campaigns have achieved success 53 percent of the time, compared with 26 percent for violent resistance campaigns." In a recent TED talk, Chenoweth said

[103] M. J. Stephan and E. Chenoweth, "Why Civil Resistance Works: The Strategic Logic of Nonviolent Conflict," *International Security* Vol. 33, No. 1 (Summer 2008): 7–44.

they have found that, "no campaigns failed once they'd achieved the active and sustained participation of just three point five percent of the population."

Several social movements have embraced the principles of nonviolence, including the emancipation of India (which, despite the nonviolence preached by Gandhi, still culminated in very violent episodes). More recently, antigovernment protests and uprisings in Libya, Syria, and Ukraine began as nonviolent movements but eventually became armed rebellions.

What I'm trying to highlight is that many violent upheavals have been the consequence of social inequality, injustice, and oppression. Peace will be achieved once we are driven by generosity, not greed. Once we're motivated by sense of duty, not indolence, cooperation instead of competition, gratitude instead of envy.

In the sixties, a Colombian priest, one of the leaders of the movement known as Golconda, argued that a more just society in which we loved each other would not be possible without social justice. Obviously, love is not what motivates oppression, he said, and the oppressed are not inclined to love their oppressors.

To explain why he took political action, the priest, René García, said, "I understood that the fundamental problem (of the country and the world) was a political problem. I began to look at the structures of capitalism and the incompatibility between capitalism and Christianity."

Garcia, as the existentialists, was protesting against the transformation of man into a thing and the dehumanization resulting from industrialization and mechanization.

# PART FIVE: Love, Love

*For there is merely bad luck in not being loved; there is misfortune in not loving. All of us, today, are dying of this misfortune.*
—Albert Camus

In the philosophical tradition of the West, Plato, in Greece, is the essential reference when we talk about love. In *The Symposium*, Plato described how a lover ascends from an instinctive desire for an individual beautiful body, or basic lust, to a state that gives rise to a more intellectual idea of love, which is in turn overcome by a transcendental vision of love (the "form of beauty"), going beyond mere sensual attraction and reciprocity and achieved through some form of revelation.

Plato's ideas about love are contemplated mainly in his Socratic dialogues: the Symposium and Lysis.

Plato considered three main categories of love: eros, philia, and agape.

Eros is that part of love that's characterized by a passionate and intense desire for something. Not purely human, nor purely divine.

Philia refers to non-passionate love, the affection and appreciation that one feels for friends, for comrades, for family, the polis (the political community), work, and, ultimately, for humanity. Philia is characterized by loyalty, service, and ideal sharing. Note that the Greek word *philia* could be translated as friendship but also as love.

Agape refers to the type of unconditional love that nourishes us. A love of a fraternal kind, compassionate, gentle, selfless, altruistic, where pleasure is found in giving. It follows the golden rule. The first Christians used this term as a synonym for *charity*.

Of course he also talked about another kind of love: philosophy, the love of wisdom, the love of knowledge.

The Greeks also considered other forms of love: ludos (love between two people, playful, without commitment), storge (family love), pragma (refers to unions that share a common interest), philautia (love for oneself). Xenia, was Greek's word for hospitality and referred mostly to the courtesy and friendship between two or more people coming from different regions.

## Love as a path

*The gifts of our biology are a potential, not a guarantee.*
—Dr. Bruce Perry

In *The Emphatic Civilization* (TarcherPerigee, 2009), Rifkin observes that, historically, empathy arises in parallel to the capacity of the human being for individuality (selfhood). The human being easily expresses and feels sympathy, which is merely the ability to feel sorry for the one who's suffering, and, therefore, it's a passive experience. Empathy, on the other hand, implies our desire to understand what the other is experiencing. Compassion, going a step further, includes the intention to share feelings and offer selfless support.

By developing the capacity for introspection and gaining the ability to explore our feelings toward others, we would begin to experience empathy, expressed as a recognition of our shared humanity: I feel your pain, I know what suffering is. Or, I might not have gone through this same experience, but I can understand or imagine how you feel. Empathy is not an emotion that can be feigned. Empathy must be authentic and can be cultivated. At the end of the day and, as we've repeated throughout the book, we are physiologically equipped to experience it. That's why, if paying attention and being authentically present, we can capture the expression of the person with whom we're interacting, sense their state of mind, resonate with their feelings, and connect with them. We use the mirror neurons for that purpose.

Let me mention that in 1949 a Canadian neuropsychologist, Donald Hebb, had already postulated a neuroplasticity theory often expressed as "cells that fire together, wire together." This theory helps explain associative learning, in which the more you use a neuronal circuit or neural pathway, the more the strength of the synapses involved increases. This is relevant to my objective to support the notion that empathy and love can be learned, and the more opportunities we have to practice, the more empathic and loving we will become.

The psychoanalyst Erich Fromm recognized the existence of different kinds of love, based on the object of our affection: fraternal (which Fromm considers part of all other types of love), maternal and paternal love, erotic love, love of self, and love of God. For Fromm, our need for love is related to the experience of separation from the mother, occurring as the result of our birth.

One of the ideas that impacted me the most when I read *The Art of Loving* as a teenager was that Fromm considers love an attitude, a character orientation that determines how a person relates to others. He said, "If a person loves only one other person and is indifferent to all others, his love is not love but a symbiotic attachment, or an enlarged egotism." In other words, if I truly love a person, I'd love all people; I'd love the world; I'd love life.

I adhere to the idea that, from a spiritual point of view, life is an evolutionary path (an adventure of the soul, the soul being that essence of ours that impels us to explore what life offers and to learn from it). Buddhists have proposed that life is a path toward enlightenment and the end of suffering. In Buddhism, it's believed that love nurtures the freedom sought. In *Ethics for the New Millennium* (Riverhead Books, 1999) the Dalai Lama said,

> Thus, we can reject everything else: religion, ideology, all received wisdom. But we cannot escape the necessity of love and compassion....
>
> This, then, is my true religion, my simple faith. In this sense, there is no need for temple or church, for mosque or synagogue, no need for complicated philosophy, doctrine or dogma. Our own heart, our own mind, is the temple. The doctrine is compassion. Love for others and respect for their rights and dignity, no matter who or what they are: ultimately these are all we need.

For Buddhists, love and compassion are just two aspects of the same thing: compassion, a desire for the other to be free of suffering, and love, a desire for the other to experience happiness.

Christians also propose that love is a way to find salvation, by following Jesus and practicing love of neighbor.

For Sufism, a mystic philosophy whose origins date back to more than fifteen centuries ago, life is a way to find our deepest identity, which transcends the apparent personality and is in harmony with everything that exists. Not very far from the teachings of Taoism, an ancient Chinese philosophy that teaches the importance of living in harmony with the whole.

But you don't need to adhere to any religion or philosophy to embark on a loving path. Life presents us with the opportunity to learn different forms of love: a learning that, I believe, takes place through a series of stages. Sometimes our learning process stagnates or deviates for individual or social reasons, and the consequence is that we disconnect from others.

My own clinical observations and a practice as a psychotherapist, Reiki practitioner, and Reiki master/teacher have allowed me to witness how much we humans crave connection and are biologically and genetically inclined to depend on others but also to support others. We are prone to care for children and the elderly, to experience positive emotions, to empathize with those going through difficult times, to share feelings with those who open their hearts to us, to cooperate with those who need support, to protect the disabled, and to be outraged by injustice.

While these dispositions may very well be scientifically explained as adaptive biological advantages, it's also interesting to explore other theories and cosmogonies that underscore a human tendency to establish a transcendental connection with something that goes beyond the biological and rational universe. Inquiry that would reveal, for example, the reasons underlying certain acts of courage that, from an evolutionary point of view, seem counterintuitive. Acts like risking one's life to save another person (even a stranger), motivated by concern, compassion, love of the motherland, or adherence to a cause. Reciprocal altruism theory tries to explain why individuals assist nongenetically related others.

Both love and compassion generate pleasant feelings in us and play a role in the ability to modulate our behavior. These feelings are related to neurobiological functions and molecular activity occurring in the circuits associated with motivation and reward in the brain's limbic system.

Science offers no clear answers about what was first: biochemistry, activation of neural circuits, behavior, or feelings. But neuroscientists have studied and proven that our ability to regulate emotions and connect socially through love and compassion definitely has a positive effect on stress management, health, life satisfaction, and, of course, our relationships.

Another vast field of study, traditionally the object of metaphysics, explores biomagnetic fields emitted by the body. Thanks to the increasing sophistication of certain measuring devices there's now evidence of the existence of those fields not only in human but in all living beings. For example, the superconducting quantum interference device (SQUID) magnetometer is capable of measuring the biomagnetic field generated by a single heartbeat, a muscle contraction, or the activity pattern of a brain cell. Even the magnetic waves surrounding the body can be measured with the SQUID, a device currently used in research in various medical centers around the world.

Traditional Chinese medicine and Ayurveda (traditional medicine from India) have proposed the existence of energy fields (known either as qi or prana) since about five thousand years ago. The belief is not only that energy nurtures our internal organs but also that health is impaired when energy can't freely flow in our physical, emotional or mental body.

Both Chinese and Hindus emphasize the importance of adopting practices aimed at cultivating, storing, and guaranteeing the flow of that energy in and around the body in order to maintain not only good physical and mental health but our connection with the whole to which we belong.

Esoteric philosophies originating in the West have postulated the existence of what they call "subtle bodies" (the physical body would be a "dense body"), each corresponding to a different plane of existence, including what in theology is known as spirit or soul: an essence going beyond what could be explained in terms of DNA, genes, or biological evolutionary advantages. These bodies would instead be linked to our thoughts and feelings, our generosity, creativity, and altruism, and to the evolution of our ability to love. Esotericism also believes that intangible bodies explain individual differences that cannot be elucidated just by taking into account the upbringing, socio-cultural experiences, or natural talents. Why are some beings kinder than others or more inclined to serve, to give, to love? But, especially, what motivates us to be better? These are the same questions scientists have asked when trying to explain consciousness, memory, and imagination.

## We, love apprentices

*No one is born hating another person because of the color of his skin, or his background, or his religion. People must learn to hate, and if they can learn to hate, they can be taught to love, for love comes more naturally to the human heart than its opposite.*

—Nelson Mandela

Many couples coming to therapy find themselves unconsciously repeating relational and communication patterns learned from their families of origin, even when they're critical of their upbringing. It's rather by chance that couples find the opportunity to live a conscious process in which they adopt their own definitions of love or find the unconscious source of their expectations of the other person.

Our biology provides the paraphernalia, the wiring, the potential, but the way we love, the language we use to express our feelings, the relationship patterns, they are all learned, not innate. The learning happens within the family of origin, the extended family, the school, and society. Is it possible for us to get rid of the unconscious conditioning and develop our love potential with full awareness?

It is unfortunate that we've grown accustomed to a reductionist, deterministic version of science. We've learned to think in terms of cause and effect instead of using systemic approaches in which all the elements that make up a system are seen as interdependent and contributing to the functioning of the whole.

We are born with billions of neurons in the brain, all with the same DNA. In the embryo, these neurons start to differentiate and specialize in diverse functions (vision, touch, movement). Changes happen – epigenetics explains– thanks to the activation and deactivation of genes, depending on the environment, and those changes affect how genes command cells to produce certain proteins. This process continues throughout our lives. So, if we were to argue that genes determine our behavior, we would need to remember that although genes command the coding of molecules that stimulate cell activity, genes only express (turn on) if certain conditions in the environment are met. That is, we are the product of nature (genes) but also of parenting, education, and the environment at large (nature). In addition, our neural circuits and brain networks are modeled after behaviors we repeat again and again.

Very often, as adults, we get used to functioning on autopilot because we have incorporated patterns of movement and thought. We get home after work without having to rethink the route we're taking. In the same way, we tend to assume that our thoughts, our beliefs, our doing are all valid. We believe we've discovered the one right way of thinking, doing, or relating. We get used to seeing the world uncritically and to behaving in such ways that we tie our perspective, beliefs, and behaviors to our identity. We become our thoughts, our performance, our titles, our possessions.

Our identity becomes entwined with our beliefs and customs, and we resist change. We filter new information, taking only what fits and confirms our identity. We feel surprised when meeting someone who doesn't share our beliefs or doesn't follow our group's socially accepted behaviors. We are biased. The others must be wrong. We tend to denigrate those who aren't like us. A self-centered perception leads us to suppose that the other must feel and think like us, that they must see the world through the same lenses we do.

Our capacity for emotional resonance, or to connect to a brother, a friend, a couple, the affective language we use, are all learned in social contexts. It happens through establishing relationships, modeling behaviors, through being lectured and advised, and, oh yes, by watching movies. But unlike with the learning of, for example, science or mathematics or language, you won't find curricula in schools or professional development programs aimed at helping you develop empathy or loving skills.

The Collaborative for Academic, Social, and Emotional Learning (CASEL) was created in 1994 to promote the following competencies in the curricula of preschool through high school education: self-awareness, self-management, social awareness, relationship skills, and responsible decision-making. They have partnered with twenty-one school districts, (of the existing 13,506 districts nationally) serving 1.8 million students.

No doubt these programs do contribute to helping students regulate emotions and work with others. Hopefully some of them will also stimulate empathy and compassion.

Teachers also need support. Their jobs are highly stressful and high rates of depression and substance abuse are reported among them.

When conflict arises between young people, teachers don't necessarily feel compelled to intervene or model the best ways to resolve conflict. And why not? If we instruct children in mathematics, biology, and geography, why not also educate them in love, mutual respect, and effective communication? They need to be prepared for caring for the planet, devising technology that contributes to peace, developing meaningful relationships, and being parents. Shouldn't we teach and test people's empathy skills and sense of responsibility before letting them undertake such responsibilities? Why leave that learning to chance?[104]

As for parents, I see that they're more likely to attend a court-ordered parenting class take the initiative to seek trainings, self-motivated to explore which practices could cater to the specific needs of their family. It's the same with couples: they don't usually attend classes to learn effective communication, active listening, synchronization, mutual respect, or even household finances—which are often the cause of conflict. (A motivated person can, however, find such instruction.)

## Step by step we go

Human development theories suggest that our intelligence, skills, and mind evolve over time. The best-known theories focus on different aspects of social, emotional, communication, motor, or cognitive development and explain the existence of a continuous process of transformation. Progress through each stage, is characterized by distinctive conflicts that need to be resolved and goals that need to be achieved. Each stage builds on preceding stages.

[104] Please explore the *MindUp* and the *Roots of Empathy* curricula.

Risking a simplification, I'll offer succinct summaries of significant and widely accepted child development theories. Then I'll propose that our capacity for love (and empathy and compassion) also develops through stages in a sequence that runs parallel and is interdependent with maturation processes taking place in the psychosocial, affective, cognitive, moral/spiritual, and language areas. Most of the current child development theories build on the following:

1. Behavioral theories of development proposed by Ivan Pavlov and B.F. Skinner. Their theories emphasize the influence of environmental factors on learning and behavior.
2. Sigmund Freud and Erik Erikson's theories of child development, considered psychoanalytic theories.
3. Arnold Gesell, known as "the father of child development."
4. John Bowlby's attachment theory, considered social developmental theory.
5. Cognitive theories that were foremostly addressed by Jean Piaget and Lev Vygotsky.
6. Martin Hoffman's studies on the social and emotional development of the child, especially in what concerns to the development of empathy and moral development.

Let me expand a little on some of them.

**The psychoanalytic theory of Sigmund Freud**: At the beginning of the twentieth century, Freud interpreted human development as the product of intrinsic and unconscious drives and motives that influence a person's behavior and perception of the world. Freud proposed the existence of five stages in human psychosexual development: oral, anal, phallic, latent, and genital. According to Freud, either the conflicts typical of each stage are resolved or a fixation in that stage occurs. A successful progression determines the attainment of psychosexual maturation; when stages are not completed successfully, he believed, our personality development stalls. Many followers have elaborated and developed Freud's theories to help us understand the quality of our attachments, our drives, our ego structure, the impact of early trauma, and the development of our sexuality. Despite the animadversion some feminists have expressed against Freud, for having postulated that women suffered from penis envy and men from castration anxiety, his theories opened the door to the sexual liberation of women. Miquel Bassols, former president of the World Association of Psychoanalysis, said that Freud was a "disgruntled misogynist, but he let himself be taught by women. Freud gave a voice to the repressed woman in the Victorian era and raised the question: what does a woman want?"

**Erik Erikson's theory of psychosocial development**: In the mid-twentieth century, the German psychologist's theory aimed to complement Freudian thought. In *The Life Cycle Completed* (Norton, 1962), Erikson said we all advance through eight stages from childhood to old age, each of them with its typical crises, whose resolution determines how we move toward the next stage and how we structure our personality. Erikson anticipated that we needed to develop a healthy identity during adolescence in order to achieve intimacy in romantic relationship during adulthood.

Pediatrician, educator and child psychologist **Arnold Gesell** postulated the theory of maturation around 1925. He believed that we are born with innate patterns or internal mechanisms, product of evolution, thanks to which we mature in a fixed sequence of six archetypal stages and that we oscillate between period of equilibrium and imbalance periodically. He considered that the cultural, social and other factors play a role in the maturation process. He helped us understand that the child isn't a small adult, immune to the events happening around him. He founded the Children's Study Center at Yale University.

**Jean Piaget's theory of cognitive development**: Toward the middle of the twentieth century, epistemologist, psychologist, and biologist Jean Piaget demonstrated that in the process of representing and reasoning about the world, humans advance from simple, symbolic, preverbal forms of thought toward logical and abstract thought, and this process occurs through identifiable stages of cognitive development that span from birth to adulthood. For Piaget, learning is an ongoing process during which the child, as the subject of her own learning, constructs new meanings, not her parents nor her teachers (as later proposed by Albert Bandura and Vygotsky in their social constructivist approach). He established that it's the individual who lays the foundations of his own learning process, dependent on how he organizes and interprets the available information. Piaget's theories contributed significantly to the understanding of children as active subjects of their learning.

**Martin L. Hoffman's theory of moral psychology**: In his book *Empathy and Moral Development: Implications for Caring and Justice* (Cambridge University Press, 2000), Hoffman identified five categories of empathic arousal in the development process of the human being, according to which, as the child's cognitive and perceptual abilities mature, it becomes easier for them to recognize the different signs of emotional distress in the other and become capable of responding with behaviors aimed at alleviating their distress, which in turn makes the child feel better.

The first three categories Hoffman described are preverbal, automatic (instinctive?), and essentially involuntary. As they precede our sense of identity, they seem to prove our predisposition to be empathic. The following are the categories Hoffman proposed:

**1. Motor mimicry and afferent feedback**: The child reacts to another child's cry by seeking the mother for comfort, as he does when he is sad.

**2. Classic conditioning**: The child offers comfort to another person.

**3. Direct association**: The child seems to take cues from the person who suffers and associates the suffering with his own past experience. The child is capable of recognizing her individuality and that of the other. Now she can progress to expressions that require a more advanced cognition.

**4. Mediated associations.**

**5. Role-play or perspective-taking.**

The last two categories require a higher cognitive functioning. The fourth category allows the observer to make associations between what the sufferer is experiencing and their past experiences. The fifth allows the observer to imagine how the affected person feels as if it were his own experience (what we call putting oneself in someone else's shoes).

Hoffman explained that we naturally respond with empathic distress to another person's suffering. According to Hoffman, a person's prosocial *moral structure* is a network made of empathic affects, cognitive representations, and motives. It's our empathy toward a potential victim or those who are helpless that impels us to help them.

Hoffman also postulated that the development of empathy is proportional to the inhibition of aggressive and violent behaviors. According to Hoffman (In H. R. Schaffer, *Social Development*. Oxford, UK: Blackwell Publishers, 1996), empathy develops through four stages:

1. ***Global empathy***. In the first year, children may match the emotions they witness (e.g., by crying when another infant is crying), but the emotion is involuntary and undifferentiated.

2. ***Egocentric empathy***. From the second year on children actively offer help. The kind of help offered is what they themselves would find comforting and is in that sense egocentric; nevertheless, the child at least responds with appropriate empathic efforts.

3. ***Empathy for another's feelings***. In the third year, with the emergence of role-taking skills, children become aware that other

people's feelings can differ from their own. Their responses to distress may thus become more appropriate to the other person's needs.

4. ***Empathy for another's life condition***. By late childhood or early adolescence children become aware that another person's feelings may not just be due to the immediate situation but stem from a more lasting life situation. Empathy may also be felt with respect to entire groups of people (the poor, the oppressed) and thus transcend immediate experience.

In the seventies, Hoffman studied the relationship between affect, cognition, and altruistic motivation and published several papers on sex differences in empathy and related behaviors. He found that empathy seemed to be more relevant in women. His conclusions have been recently validated by studies in animals and infants that seem to provide phylogenetic and ontogenetic explanations confirming that sex differences in empathy go beyond cultural expectations of gender roles.

John C. Gibbs, author of *Moral Development & Reality: Beyond the Theories of Kohlberg, Hoffman and Haidt,* (NY: Oxford University Press, 2019) enriches Hoffman ideas with recent research, while providing insights of his own.

## Nurture vs. nature?

Human development theories have not been alien to the old debate about whether we are a product of nature (genes, natural selection) or of nurturing (environment) or both.

In the last century the English psychiatrist and psychoanalyst John Bowlby questioned Freud's theories of development, arguing that while psychoanalysts scrutinized the inner, unconscious world of a client, they failed to explore the influence of the parents on mental structures and development. Freud had originally proposed that the neuroses he observed in his practice resulted from actual traumatic events (especially of a sexual nature) that had occurred in the patient's childhood. However, he seems to have eventually abandoned this idea in favor of one involving infant *phantasies* (Freud's term for states of the mind stemming from genetic needs, drives, and instincts), which he considered part of a normal human's psychical structure (Oedipal complex, envy of the penis, castration anxiety). These, Freud claimed, were associated with neurotic symptoms.

Instead, Bowlby (*A Secure Base: Parent-Child Attachment and Healthy Human Development,* Basic Books, 1990) was especially interested in the attachment between child and caregiver. He concluded that human beings tended to form bonds because we can't survive without others. He went even further, suggesting that our interaction with parents and other caregivers in the early stages of our lives determines the quality of any relationships we establish in the future.

After Bowlby, Canadian psychologist Mary Ainsworth found a way to classify and determine the quality of the mother-child infant attachment to predict some features of child development that would manifest in adult life. Ainsworth's observations made it possible to demonstrate that, for the child's attachment to her mother figure to be safe, the caregiver must be able to perceive, understand, and respond appropriately to the signals the child gives at each moment. She believed that the emotional security of the child, who must feel she's unconditionally accepted and protected, determines the quality of the attachment.

Some post-Freudians are critic of the attachment theory, arguing that there's a difference between the concepts "nurture" and "environment," terms that have been often used as synonyms. According to some Freudians, both Bowlby and Ainsworth put a great burden on the parents by focusing on the importance of nurturing, while leaving aside the influence of peers and the culture (environment).

Daniel Siegel at the University of California at Los Angeles (UCLA) and the Mindsight Institute, has further studied how the brain and the mind are shaped from our interaction with each other and the environment. He emphasizes on how relationships with caregivers shape our mental universe. In *The Neurobiology of We* (Sounds True, 2008), Siegel explores in detail how the quality and particularities of the attachment a child develops with the mother figure have an impact on the subsequent psychosocial development of the child.

## Take a look at alternative perspectives

*The mystic and the physicist arrive at the same conclusion; one starting from the inner realm, the other from the outer world.*
—Fritjof Capra

It would not be fair to end this segment without mentioning a few paradigms of human development, including those typical of the philosophies of Eastern countries. The main idea across these different beliefs is also that our lives are organized along a series of goals and tasks that we must accomplish in order to either become enlightened or develop full potential.

In the fifties, Abraham Maslow[105] proposed a theory of human motivation, including a hierarchy of needs that should be satisfied in a certain order. Starting with the most basic needs—those that guarantee survival—we'd advance step-by-step toward self-realization. When the needs in one of those hierarchies have been more or less satisfied, our activities can be directed toward the satisfaction of the next group of needs. For example, once physiological and safety needs are satisfied, we can proceed to meet the social needs of loving and belonging, which would explain why we form families, why so many people seek to become members of a congregation or join a political party, a club, or a sports team.

Because Maslow didn't offer empirical evidence of his theory, psychologists sometimes disregard his linear model. In 1969, Maslow

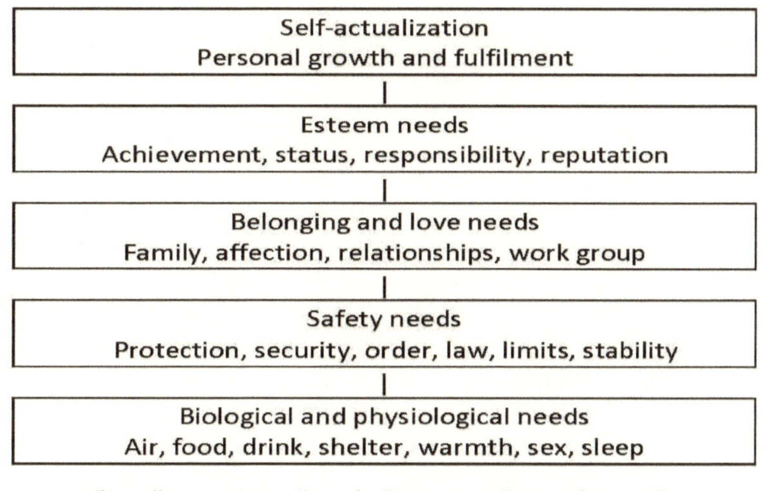


*Abraham Maslow's hierarchy of needs*

established the Association for Transpersonal Psychology. For more than a century, Transpersonal Psychology has embraced the idea that spiritual and transcendent aspects of the human experience need to be contemplated in any study of human development.

Almost all human cultures have developed a concept of mind, spirit, or soul as something discernable from the physical body. The idea of an

---

[105] Abraham Maslow, Robert D. Frager (Editor), James Fadiman (Editor) "Motivation and Personality." (New York: Harper and Broth, 1954.)

ethereal vital principle animating the life of humans and other beings, is also common to many religions. By the time of Socrates's death (399 BC), the Greeks seemed to have a unified thought about the soul as an essence that characterized all living things. In humans, that essence was the psyche, where our thoughts and emotions resided, adorned by virtues or corrupted by vices. But the most elaborated ideas are found in the oldest traditions, such as Vedic (fifteen to five centuries BC), Buddhism (five centuries BC), and also in some branches of traditional Chinese medicine (twenty-five to five centuries BC). These philosophies not only postulate the existence of psychic-energy centers, known to Vedas and Buddhists as chakras, and to Chinese as dantians, but also teach that these centers undergo a process of development throughout life in what I consider another significant and relevant lifespan developmental framework.

Hinduism, Buddhism, Taoism, and Jainism share the belief that in addition to our physical body, made of what they call dense energy, we have subtle, energy bodies: etheric, mental, emotional, spiritual bodies, each one of them having a characteristic vibrational frequency that manifests as colored lights, from the lower frequency (red, physical) to the highest frequency (white or violet, nirvanic or monadic).

Based on the Vedic tradition, the late spiritual guru Osho explained that at birth, our physical dimension is the only one that is already developed (we're born with a whole visible physical body), and each of the other bodies or dimensions exist only as a "seed." It would take about seven years for each one of the other dimensions to develop with their own characteristics. As each body develops, thousands of possibilities open for us, depending on our conscious effort and the kind of life we're following.

As for the development of the lower, middle, and upper dantians, they are believed to be, respectively, essential to achieving will power, concentration, and spiritual connection.

Yogis say we evolve spiritually by focusing our awareness on the soul to eventually achieve coherence between action, word, and thought. This coherence would allow for the maturation of each one of our energy bodies. Not everybody would necessarily achieve maturation of all seven "subtle bodies." As with all other skills, this development requires discipline and work.

Some Hindu traditions believe in the existence of up to eighty-eight thousand chakras[106] connected by channels called nadis. Others believe that each individual has a different number of energy centers. But most

---

[106] C.K. Best, (2010). "A chakra system model of lifespan development." *International Journal of Transpersonal Studies*, 29(2), (September 27, 2010). http://dx.doi.org/10.24972/ijts.2010.29.2.11

seem to agree that there are at least seven main centers located in these areas: the perineum, sacral region, solar plexus, heart, throat, third eye, and crown. Each chakra is seen as an energy center correlated with physical, psychic, and spiritual functions (and corresponding with each one of our energy bodies). These centers would gather information/energy and distribute it to keep our body-mind running smoothly. It's believed that the conscious development of the chakras facilitates our advancement in the path to enlightenment. As access points, chakras allow the flow of essential vital energy to nurture our inner organs. The development of each chakra is also associated with an evolutionary task.

The first chakra, located on the perineum, regulates the instinct for survival and our connection with the earth and our roots (family, tribe). It is believed to contain a mysterious divine potency—kundalini, or energy of consciousness—which, if cultivated, can ascend to our seventh center and lead us to enlightenment (or connection with the ultimate). The second chakra (*swadhisthana: the dwelling of self*) regulates our attraction and love for others (sexuality, sensuality, empathy) and the third, our personal power (self-esteem, willpower, control of emotions, and ability to tolerate frustration). The three lower chakras correlate with the denser bodies (physical, emotional, and mental).

The chakras must develop and open throughout our lives if we are to achieve a state of harmony within ourselves and with the world and, eventually, enlightenment. For this purpose, the lower frequencies must be harmonized with the frequencies of the upper chakras. When the fourth chakra, at the level of the heart, develops and opens up, a person begins to experience increased compassion and an inclination to service, which facilitates the opening of the fifth chakra (located in front of the throat), allowing us to evolve toward a greater creativity and a call to teach. The development of a greater intuition would depend on the opening and development of chakra number six, in the area of the third eye, between our eyebrows, and a sense of transcendence (loss of self)

will be achieved with the opening of chakra number seven, on the top of the head.

The Austrian philosopher Rudolf Steiner, founder of a spiritual movement known as anthroposophy, also considered this system of chakras as a dynamic whole that evolved with man, and he applied it to his educational philosophy. I visited a Waldorf school in Carbondale, Colorado, a few years ago. It follows Steiner's esoteric philosophy, applying it even to the way their facilities are built and the decoration of the classrooms. "Color is the soul of nature…and when we experience color we participate in this soul," Steiner said. Classrooms were painted according to the stages of spiritual development that he described. His ideas of development were not very far from those of Piaget.

Mention of the chakras is found in the Vedas, Indian ancient writings (1,700–1,000 BC). They're also mentioned in the Yoga Sutras of Patanjali (compiled prior to 400 AD, who synthesized Hindu's wisdom. Sir John George Woodroffe translated and published, under the pen name of Arthur Avalon, several ancient texts. We owe him for the knowledge the Western world has about the Vedas as a noble and ethical philosophical system.

Another developmental paradigm worth mentioning comes from the Kabbalah: it's known as the *tree of life* and is a symbolic map for the forces in the universe and their reflection or correspondence in the human body. It consists of ten *Sephiroth* connected by 22 paths. Each sephira represents successive divine emanations that correspond to ten stages in the continuous evolution of the universe, man and things manifested. Each Sephira is a seed that contains certain potential. The goal of a cabalist is to transform this potential energy so that its free flow creates balance in our lives.

# PART SIX: A Path Leading to Altruism

*In the development of mankind as a whole, just as in individuals, love alone acts as the civilizing factor in the sense that it brings a change from egoism to altruism.*
—Sigmund Freud

I propose that our capacity to love, experience empathy, feel compassion, and cooperate, needs to be examined from a developmental perspective. To fully develop our ability to love, we must advance through stages, completing specific developmental tasks, before we can undertake new learning challenges.

Although some researchers debate weather self-love is a prerequisite to developing the ability to love others, there is some evidence that learning self-compassion makes us more compassionate toward others (there's also some evidence that people tend to be more compassionate to others than to themselves).[107] In contrast, narcissistic love doesn't leave any libidinal energy available to invest in loving others, and a low self-esteem would hardly facilitate loving others (thought it may make one want to please others and behave submissively, which could create the appearance of a loving predisposition).

I propose that once the early stages in the development of empathy and love progress successfully, the child can move through subsequent stages to become an adult with a healthy capacity to fully love self and others and find joy in life.

I also propose that as we progress through the stages of social, cognitive, communication, and psychosexual development, we also experience the opportunity to learn and develop different types of love. And along the process of fulfilling our personal emotional needs and responding to socialization demands, we develop empathy and compassion, which are requirements for fully loving others and being of service. As we successfully progress through each stage, we become

[107] A. Lopez, R. Sanderman, A. V. Ranchor, and M. J. Shroevers, "Compassion for Others and Self-Compassion: Levels, Correlates, and Relationship with Psychological Well-Being," *Mindfulness*, Issue 1 Vol 9, (Feb. 2019): 325–331.

increasingly capable of experiencing impersonal and unconditional types of love. Since they have a genetic component, each stage requires us to successfully finish typical tasks and face specific risks before we can progress to the next stage.

We know that not all love is of the same type. Some forms of love are personal, others impersonal. Some, such as maternal love and fraternal love, or love for the planet, contain elements of unconditionality. In these cases, we offer love without expecting reciprocity. Other forms of love are conditional, like the love of the little child toward his caregiver. Love can also be conditional in some cases, as with couples, where certain requirements must be met for a person to stay in a relationship. In committed, long-term relationships, however, love tends to be unconditional.

We already talked about the most salient developmental theories. Now I'd also like to share a few thoughts about adult growth and development.

The psychologist Clare W. Graves dedicated the last twenty years of his life to formulating a theory of adult development centering on his belief that "the psychology of the mature human being is an unfolding, emergent, oscillating, spiraling process, marked by progressive subordination of older, lower-order behavior systems to newer, higher-order systems as man's existential problems change." In other words, learning and growth are continuous (don't end when we become adults) and seem to happen in an upward spiral (more than in a linear or vertical progression) of increasing complexity. According to Gravesian theory, we meet learning challenges repeatedly but each time at a higher order of complexity, conquering them once we acquire the needed new knowledge, wisdom, and expanded perception.

Now I propose that our ability to love evolves *across* the whole lifespan and that this ability could and should be stimulated.

Applied to the progress of learning to love ourselves, the spiral model would mean that we're born self-invested, we soon learn to love others, and as we practice and improve self-love, self-esteem, and self-compassion, we would find new opportunities for growing and for loving ourselves and others at different stages of our lives.

What follows are rough ideas about the individual evolution of love. They are part of a work in progress and require further elaboration and a cultural context, since they might not be universally valid. Children in Eastern and Western cultures are raised differently. For example, most Western children are socialized toward psychological autonomy, but non-Western children in rural areas are more likely to be socialized toward relatedness, which would have a different impact on the development of empathy, compassion, and love. Drawing from psychoanalytic and other development theories, I propose that, in ideal situations, our ability to love develops according to the following stages and that each one is the foundation for the development of the next:

**Self-love—egocentric love**: This stage begins at birth and sets the foundation for self-esteem. Some psychoanalytic currents refer to this first stage as primary narcissism. At birth, the child is not yet self-aware, and the external world only makes sense to the extent that it satisfies the child's needs. Children become invested in exploring and discovering their bodies. They rejoice in their own charms and feel loved for what they are. Erik Fromm says, "Mother loves the newborn infant because it is her child, not because the child has fulfilled any specific condition, or lived up to any specific expectation."

Babies see themselves in their relationships with their mothers and other close people, who function as mirrors[108] to help them know themselves. This stage seems to extend from birth to six or eight months, coinciding with what Margaret Mahler called the symbiotic stage.

**Love for the mother figure**: Ideally, the mother (or mother figure) is a source of gratification, and little by little, the child recognizes her as otherness, attachments develop, and she becomes a bridge to the outside world. This is a primary bond that the child forms and maintains with the mother. According to Daniel Siegel,[109] the formation of a "secure attachment" predicts a healthy mental development, including a capacity for healthy interpersonal relationships. In this stage, the child builds internal maps of self and others. This stage appears to go from six or eight months to about three years, coinciding with the Freudian oral and anal stages, and Margaret Mahler's separation-individuation stage.

**Love for a third person**: After establishing a secure attachment with the mother or mother figure, the child gets increasingly curious about the world and forms a bond with a father figure or another person who takes care of the child. The active exploration of the world begins. The inclusion of a third person allows the development of sharing, learning to take turns, understanding limits and boundaries. This lasts from six months to school age, coinciding with what in psychoanalysis are known as the oral, anal, and phallic stages.

---

[108] D. W. Winnicott, "Playing & Reality," Chapter 9, *Mirror-Role of Mother and Family in Child Development* (London: Tavistock, 1971).

[109] D. Siegel, "*The Neurobiology of 'We': How Relationships, the Mind, and the Brain Interact to Shape Who We Are*" (New York: Norton Series Audiobook, 2008).

| Love Development | Learning and task | Achievement | Cognitive stage Piaget | Psycosocial -sexual Freud Erikson | Risk | Empathy Hoffman |
|---|---|---|---|---|---|---|
| **Self love 0-2 yrs Personal/ inconditional** | *Integration of body image Formation of a sense of self/ego* | *Self-esteem Security* | *Sensorimotor initial Pre-verbal* | *Oral stage Trust vs. Mistrust* | *Narcissism* | *Global* |
| **Mother love 6-36 mo**<br><br>**Personal/ Conditional** | *Self-care, attachment with mother figure Symbiosis. Awaress of the world out there that nourishes me, of which I depend* | *Separation & individua-tion. Formation of ego Identity.* | *Recognizes external world - Interact Language without communicative function. Symbolic, egocentric thinking* | *Annal stage Autonomy vs. Shame* | *Dependency* | *Global conditional egocentric* |
| **Third person love 18m – 5 yr Personal/ Conditional** | *Love for third person, inclussion (father) Primordial group. Triangulation. Integration to society. Law* | *Order Loyalty Inclussion* | *Improvement of language. Operational concrete thinking* | *Phalic stage Edipal Iniciative vs. guilt* | *Conflict* | *Egocentric* |
| **Fraternal love conditional filial 3 a - puberty**<br><br>**Personal/ Unconditional** | *Personal love and beginnings of unconditional love. Belonging, social interaction. Projection, Acceptance, Solidarity Negotiation* | *Sharing - conditional being in community – Genderidentity conditional social* | *Concrete thinking to operational (abstract) formal* | *Latency stage Pre-puberty Puberty Industry vs. inferiority* | *Lack of acceptance Intolerance* | *For the other person's feelings* |
| **Couple's love Adolescence- Young adult Personal/ unconditional** | *Personal love Erotic love Fellowship Sharing, loyalty* | *Assertion Autonomy Interdependency* | *Abstract formal thought* | *Identity vs. Confussion* | *Relationship failure Narcissism Grief for losses* | *For the other person's feelings Cognitive empathy* |

| Love Development | Learning and task | Achievement | Cognitive stage Piaget | Psycosocial -sexual Freud Erikson | Risk | Empathy Hoffman |
| --- | --- | --- | --- | --- | --- | --- |
| **Love for children Young adult Personal/ Unconditional** | *Unconditional personal love. Generosity, appreciation,* | *Nurture, teach* | *Abstract formal thought* | *Intimacy vs isolation* | *Parenting failure* | *For the other person's feelings & life conditions* |
| **Love for society /humanity Adolescent and up Impersonal/ Unconditional** | *Impersonal, unconditional love Generosity, productivity, creativity, detachment* | *Nourish, educate, serve* | *Formal thinking/ altruistic thinking* | *Generativityd vs stagnation* | *Self-denial* | *For the other person's feelings & life conditions* |
| **Transcendent love (Metaphysical adolescent) - Adult senile Impersonal /unconditional** | *En adolescente: poder personal. Amor a sí mismo conditional amor impersonal. Yo si puedo. En adulto: Balance, vida rítmica, sensación de plenitud conditional logro. Conexión conditional desprendimiento* | *In teen years: personal power. Self-love – conditional impersonal love. I can. In adult: Balance, rhythmic life, conditional. Fullfilment. Conditional connection, detachment* | *Formal thinking/ altruistic thinking* | *Integrity vs despair* | *Declining cognitive, isolation, depression (grief for loses)* | *For the other person's feelings & life conditions* |

**Fraternal and filial love**: The child's inclusion in the group (family or other) with a particular identity and role is established. The central tasks of this stage are achieving psychological autonomy and a foundational identity (gender, family, nationality). This stage develops through school age, puberty, and adolescence.

**Love for a partner**: There is a search for affiliation, gratification, physical pleasure, companionship, validation, and respect. The capacity to fall in love and be in love with a partner usually awakens during puberty and develops throughout adolescence and early adulthood. Typically, in its early stages, gender identity, gender expression, and sexual preference are established.

This stage usually starts with sexual experimentation during adolescence and evolves toward a committed one-to-one relationship. In her podcast series, *Where Should We Begin*, the psychotherapist Esther Perel discusses how in modern times this stage of erotic exploration has been prolonged. She said, "The moratorium period today has extended from about fifteen years of sexual and relational nomadism," and these days people enter into committed relationships later in life (https://www.estherperel.com/podcast).

Once people enter a committed relationship, this could lead to the development of a type of unconditional, personal love, which can expand over time. From the evolutionary point of view, the function of creating bonds with a partner has the purpose of guaranteeing survival.

**Love for offspring or pupils**: Typically, this is an unconditional type of love, but of a personal nature. We use the term unconditional love to describe an affection that is genuine, characterized by acceptance and empathy, and doesn't demand or expect reciprocity. This is an ideal type of love between children and parents or grandparents, but it's not necessarily real. Many parents use parenting strategies that include withdrawing affection or support from a child who doesn't behave or live up to their expectations, for example. But also, some children might significantly distance themselves from an abusive parent or from one who crushes them, not allowing them to become who they want to be.

**Love for our tribe/humanity**: At some point in life, we experience the need to serve others, to participate in political endeavors, to embrace a cause. This aspect of unconditional, impersonal love usually appears during adolescence. It's expressed as love for the planet, an eagerness to be of service or develop creative and socially productive work. It belongs to a stage in which we begin to purposefully make contributions to the world; we adopt causes, we get creative, we participate in social endeavors.

**Transcendental love**: Marked by a need to explore the meaning and purpose of existence and develop spirituality (not necessarily religiosity), this stage involves recognizing that we're all connected. It's called transcendental because it often involves experiences that go beyond our human comprehension. It germinates during adolescence, when individuals pursue autonomy, explore their ultimate identity, and embrace altruistic causes. Because of survival priorities, development of this type of love may pause during productive and child-rearing years and reappear with the arrival of maturity. It does include a form of self-love (we seek to leave a legacy behind), but also a form of unconditional, impersonal love, that involves empathy and compassion for the others and the planet. We'll expand on this later.

## Loving thyself—an adventure

*Selfishness and self-love, far from being identical, are actually opposites.*

—Erich Fromm

Psychoanalytic theory has postulated that at birth our libidinal energy is directed to ourselves. Does that mean we love ourselves from the beginning and then the environment plays a role in destroying or boosting self-love?

What determines our capacity to love self and others?

Why is narcissism, an exaggerated form of self-love, exponentially increasing in the Western world?

Disclosure: this book does not attempt to fully answer those questions.

In the twentieth century, Sigmund Freud's sequential age-and-stage developmental process theory became the main frame of reference with which to approach the study of the child's progress. Drawing from what he had learned from the analysis of his adult patients, Freud established that child development goes through five psychosexual stages: oral, anal, phallic, latency, and genital, according to which erogenous zone the child's libido is invested on.

According to Freud, one doesn't yet have an ego at birth, and all libidinal energy is directed to oneself. A child is driven by the *id*, which instinctually seeks instant gratification. A sense of self evolves in children as they feel their love is returned. They start fulfilling expectations of their *ego ideal* and gradually lose an initial sense of omnipotence (primary narcissism). Parental expectations influence the formation of an (unattainable) *ego ideal* (inner idea of who we want to become).

The Hungarian psychoanalyst Margaret Mahler observed mother-child interactions within the first three years of life. She concluded that the psychological birth of the child does not occur at the moment of birth but coincides with the appearance of the social smile (between two and three months), which contributes to the formation of a special bond between mother and child.[110]

In agreement with Freud's ideas, Mahler said that during the first few weeks after birth, the baby is totally invested in itself. (Could we call this love for self?) Afterward, the infant would develop what Mahler called a symbiotic relationship with its mother and perceive itself as one with her. The process of differentiation would begin around six months of age, once the child becomes conscious of the boundaries between self and others.

Mahler described in detail the child's process of separation and individuation that takes place during the first three years of life. She went on to postulate that if the child didn't successfully complete the process of psychological separation from the mother (or mother figure) during those early years, the subsequent stages of development could be hindered. Mahler believed that the separation/individuation process determines the development of either a healthy narcissism (self-esteem) or a pathological narcissism.

Newborns can only focus their eyes about eight to twelve inches from their face, and they see in black and white and some shades of gray. It's not until they're five or six months of age that they're able to see all colors. This perceptual fact explains in part why the baby is focused on itself.

Focusing on adolescence, psychoanalyst Peter Blos[111] later complemented Mahler's theory on the process of separation and individuation and an individual's search for identity. He postulated that childhood doesn't fully end until the adolescent manages to complete early childhood's pending developmental tasks.

Blos's contributions support the spiral model of incremental development in which we will face similar evolutionary tasks once and again throughout our lives.

---

[110] M. Mahler, "Symbiosis and Individuation: The Psychological Birth of the Human Infant, in Harrison and McDermott, *New Directions in Children Psychopathology*, Pages 89-106, (New York: International Universities Press, Inc., 1980).

[111] P. Blos, *The Adolescent Passage*, (New York: International Universities Press, Inc., (1979).

Psychoanalyst Hanna Segal[112] (object-relation theory) said that during the first months of life, the child perceives the mother as an object for its own satisfaction. "Object," in psychoanalytic terms, refers to what is different from the subject, the other, that in which the subject focuses energy (libido). Until the ego begins to exist as an internal organizer, the mother keeps playing the role of a mediator between the environment and the infant's inner impulses. At first, Segal said, the baby has no awareness of being a total person. But as the child explores the mother's body, and compares it to other perceptions, it becomes capable of discriminating between Mom and other external objects. Then the infant is ready to begin a process of differentiation (between the ages of five to ten months), progressively becoming able to discriminate between self and non-self (or me and not me).

"The infant begins to recognize his mother not as a collection of anatomical parts, breasts that feed him, hands that tend him, eyes that smile or frighten, but as a whole person with an independent existence of her own, who is the source of both his good and his bad experiences," writes Segal.[113]

To serve the purpose of shielding the child from fears experienced during the stage of individuation-separation, the infant develops what is known in classic psychoanalytic theory as a primary narcissism, which will eventually give place to a healthy self-esteem.

When he described the four stages of intellectual development, Jean Piaget observed that during the first stage (sensorimotor), which coincides in time with the oral stage described by psychoanalysis (first eighteen months of life), children typically display logical egocentrism (a term used without moral connotations). In his book *Child Conception of the World* (Routledge, 1997), Piaget said, "The child sees everything from his own point of view...because he believes all the world to think like himself. He has not yet discovered the multiplicity of possible perspectives and remains blind to all but his own as if that were the only one possible." He was referring to the initial difficulty children have understanding the point of view of another. As he interacts with the environment and grows and learns, his cognitive progress allows the child to gradually develop thoughts and feelings that include those of others.

Later, during the period Jean Piaget called preoperational (from two to seven years of age), children are still egocentric, but with regard to their representations (figurative aspects of intelligence), they assume that others

---

[112] H. Segal, *The Work of Hanna Segal: A Kleinian Approach to Clinical Practice*, (Norvale: Jason Aronson, Inc., 1981).

[113] Ibidem.

see, hear, and feel the same as they do. Even their communication is egocentric, and their thinking is more subjective than objective.

Piaget foresaw that children needed to achieve a certain degree of cognitive development before experiencing any degree of empathy (at around seven years of age, he thought). Even though more recent studies show that infants already show signs of being empathetic in the first year of life, Piaget's observations were validated by Hoffman who observed significant milestone in the development of empathy happening around age seven, when a child would empathize with the situation of another and not only with their immediate circumstances.

Researchers have noticed that newborns already show empathic resonance: they cry in response to other babies' wailing (what Martin Hoffman called "global empathy"). Ronit Roth and colleagues also found empathic responses in infants who witnessed their mothers in distress.[114]

Hoffman observed that around the first year of life, infants look for their mothers to comfort other children who are alone and crying, and around three years of age, they're able to distinguish between their own feelings and those of another child.

During adolescence, another form of egocentricity (operative) appears, which Piaget called metaphysical: the adolescent no longer feels lesser but equal to the adult and wants to outshine and surprise them by transforming the world. Grandeur and altruism emerging simultaneously.

It feels like the adolescent could see the world with a double pair of eyes: still with the same innocent eyes of a child whose thoughts transcend reality and dream, and with eyes that can simultaneously visualize what's possible. As abstract thinking develops, adolescents become increasingly aware of their ability to anticipate and interpret their experiences, and, if they so choose, they can stand peer pressure and accomplish great things.

Piaget[115] proposed that the movement from one form of egocentrism to another is dialectical, "such that the mental structures which free the child from a lower form of egocentrism are the same structures which ensnare him in a higher form of egocentrism."

---

[114] R. Roth-Hanania, M. Davidov, and C. Zahn-Waxler, "Empathy Development from 8 to 16 Months: Early Signs of Concern for Others," *Infant Behavior and Development*, Vol. 34, Issue 3 (June 2011): 447–458.

[115] D. Elkind, "Egocentrism in Adolescence," *Child Development*, Vol. 38, No. 4 (December 1967): 1025–1034.

In *All Grown Up and No Place to Go: Teenagers in Crisis* (Perseus, 1998), psychologist and educator David Elkind further explored Piaget's theories but focused on adolescents' egocentrism, which eventually leads to self-awareness. He said that as abstract thinking develops, adolescents tend to build an "imaginary audience" (building hypotheses that they test against reality) and a "personal fable" (a belief that they're special or unique). The risk of this is that it could motivate reckless behavior and feelings of being invincible. Over time, empathy and compassion will develop, given a favorable environment.

It would be difficult to grow in love and compassion without also developing a healthy self-esteem. This includes being aware that we are loved and capable (of overcoming difficulties, of purposefully achieving goals, of sacrifice, etc.). An exact opposite result should be expected when self-esteem is based on unmerited praise. (The child makes a doodle on a piece of paper and we exclaim, "How wonderful. You're an artist!")

Repeated praise of inherent abilities or for performing tasks that should be recognized as fulfillment of responsibilities, or persistent comparison and competition, stimulate not self-esteem but narcissism, creating substantial anxiety in both parents and children.

An atmosphere of constant unmerited praise has become popular. Studies show that praising the character of the child ("you're the best," or, "nobody draws better than you") could even be counterproductive.[116] When the child can't fulfill expectations, the result is shame. However, encouragement through descriptive appreciation of achievements ("I see that you made an extra effort to accomplish this," or, "I see how you focus on your task to get a good result") often renders a stimulating effect.

In "The Analysis of the Self" (International Universities, 1970), psychoanalyst Heinz Kohut considered narcissism to be the opposite of low self-esteem. According to him, a failure of the parents to display empathy and stimulate a child's sense of self could lead to both low self-esteem and personality disorders, such as narcissism, in which an individual relies on others to regulate their self-esteem and provide them with a sense of value.[117]

Kohut believed empathy played a key role in the therapeutic relationship. And he advocated for more empathy in a world in which power struggles had led to the existential threat of nuclear war. He visualized an interdependent world as necessary for humanity to survive.

---

[116] R. Vitelli, "In Praise of Children," *Psychology Today* (April 07, 2014). Retrieved online psychologytoday.com (September 30, 2019).

[117] To know more about Kohut, visit the International Association for Psychoanalytic Self Psychology portal: https://iapsp.org/kohut/.

A recent study by Eddie Brummelman and colleagues[118] shows a correlation between parents' overestimation of children as special and more entitled than others, and the further development of narcissism. As we mentioned before, narcissism is more frequent in Western countries, and not only is it on the increase, but it has contributed significantly to a lamentable upsurge in the number of violent incidents, racism, nationalism, and segregation.

The same study shows that early interventions that modify the interaction between parents and children change the course of the development of narcissism. Overvaluation doesn't show a positive correlation with self-esteem. In contrast, warmer parents, those who know how to express affection and appreciation for the child, have a more positive influence on the development of their child's self-esteem.

In another study, the same team[119] tested a self-deflation hypothesis versus a self-inflation hypothesis that links praise to narcissism and self-esteem. They found that inflated praise, which sets unattainable standards for children, actually lowers self-esteem (confirming the self-deflation hypothesis). Inflated praise also predicted narcissism but only in children with high self-esteem (partially confirming the self-inflation hypothesis).

Parents must consider the possibility that instead of praising the child, they could offer encouragement, providing feedback on their choices and on progress resulting from their efforts. It's important to be careful with messages such as "you can achieve anything you set your mind to." Even though it seems to convey faith in the child, the prompt is quite misleading. It isn't actually true that we can achieve everything we set our minds to and we better have tolerance to frustration and the capacity to accept that.

Self-awareness nurtures love for self. It involves the capacity to assess ourselves objectively, be mindful of our vulnerabilities and strengths, and discern between what we're capable of and what's beyond our reach. By adopting reasonable goals, we'll be more motivated to attain them. Each achievement would nourish our self-confidence.

A low self-esteem might mean trouble in accepting ourselves as we are. Therefore, to achieve self-love we must learn self-compassion.

---

[118] E. Brummelman, S. Thomaes, S. A., Nelemands, B., Orobio de Castro, G. Overbeek, and B. J. Bushman, "Origins of Narcissism in Children," *PNAS*, 112 (12) (March 9, 2015): 3659–3662.

[119] E. Brummelman, S. Thomaes, S. A. Nelemands, and B. Orobio de Castro. "When Parents' Praise Inflates, Children's Self-Esteem Deflates," *Child Development*, Volume 88, Issue 6, November/December 2017, 1799-1809.

According to the psychologist Kristin D. Neff from the University of Texas,

> Self-compassion involves being kind to ourselves when life goes awry or we notice something about ourselves we don't like, rather than being cold or harshly self-critical. It recognizes that the human condition is imperfect, so that we feel connected to others when we fail or suffer rather than feeling separate or isolated. It also involves mindfulness—the recognition and non-judgmental acceptance of painful emotions as they arise in the present moment. Rather than suppressing our pain or else making it into an exaggerated personal soap opera, we see ourselves and our situation clearly (quoted from her portal self-compassion.org).

Neff and Paul Gilbert and his collaborators in England[120] have found that when faced with difficulties, self-compassionate and friendly individuals show more resilience than those who severely judge and condemn themselves.

Gilbert and his team also found that developing self-compassion has a favorable effect on mood. In contrast, individuals who are overly self-critical may find it difficult to feel compassion for self or others. Another of their findings is that fear of self-compassion was linked to fear of feeling compassion for others and also to self-deprecating behavior, insecure attachments, depression, anxiety, and high levels of stress. Moreover, they found that self-criticism predicted depression. [121]

"All friendly feelings toward others come from the friendly feelings a person has for himself," Aristotle said twenty-five centuries ago.

Having self-compassion includes being benevolent with ourselves when we fail, perceiving our inadequacies as humans, and being aware of our negative qualities. It reduces fear of failure, self-doubt, and denial of one's own competences or limitations.

Based on the teachings of Buddhist scholars and her own research, Neff[122] believes that, in the process of loving oneself, it's important to

---

[120] K. D. Neff and C. Germer, Seppala, E. M., Simon-Thomas, E. Brown, S. L. Worline, M.C., Cameron C. D, and Doty, J.R. eds. *The Oxford Handbook of Compassion Science*. New York, NY: Oxford University Press, 2017, chapter 27).

[121] P. Gilbert, K. Ewan, M. Matos, and A. Rivis, "Fears of Compassion: Development of Three Self-Report Measures," *Psychology and Psychotherapy: Theory, Research and Practice*, (The British Psychological Society, 2010): 239–255.

[122] K. Neff, *Self-Compassion: The Proven Power of Being Kind to Yourself*, (New York: HarperCollins, 2011).

develop a healthy self-esteem, but to achieve it one should first be self-compassionate. In *Self-Compassion: The Proven Power of Being Kind to Yourself* (HarperCollins, 2011), Neff lists three elements of self-compassion:

1. Self-kindness vs. self-judgment
2. Common humanity vs. isolation
3. Attention vs. identification

Being aware of the present moment prevents denial of our imperfections and the repetitive reflection on the aspects of ourselves that we do not like.

Neff said, "The very definition of being 'human' means that one is mortal, vulnerable and imperfect. Therefore, self-compassion involves recognizing that suffering and personal inadequacy is part of the shared human experience—something that we all go through rather than being something that happens to 'me' alone."

## Mirror screens

Having an objective appreciation of who we are or, in other words, having a healthy self-esteem, allows us to set clear boundaries in our relationships, take a stand for our rights, create emotional distance from someone who hurts us systematically, negotiate terms in a relationship, and speak up when a situation so demands.

In contrast, when a person has developed what Delroy L. Paulhus and Kevin M. Williams called, in 2002, the dark triad traits—a combination of three negative personality qualities: narcissism, psychopathy, and Machiavellianism—their sense of entitlement makes it difficult for them to accept boundaries. They don't feel empathy or remorse. Instead of being able to negotiate terms, they're more likely to behave irrationally, both in personal and in business relationships, and to choose to speak with arrogance and impertinence when things are not going their way. That's the case for a violent partner or a sexual harasser, for example. At the other end, people who've learned to be submissive tend to yield their power and will probably lack the capacity to assert themselves. They might have trouble relating to others on an equal footing, often tolerating abuse. They feel inadequate and powerless.

It's important not to mistake humility for submission. We use the adjective *humble* more often to describe a person living or behaving modestly than to refer to a person who has developed the quality of humility, a moral virtue that allows an individual to recognize their own strengths and abilities, but without needing to be boastful about them.

A humble person has healthy self-esteem, is aware of his or her limitations, and doesn't need to compete in order to prove her value or assert power. A humble person tends to be open-minded (not arrogant), displays intellectual modesty (rather than vanity), isn't afraid to accept and rectify mistakes, and has a sense of commitment to life, which explains why they rarely get bored. But being humble in a relationship is beneficial only when we have an equally humble partner.

We already mentioned that there is evidence that narcissists are incapable of spontaneously feeling empathy. However, some researchers have found that the capacity for empathy might be also compromised in submissive people. People pleasers have an inaccurate appreciation of who they are and why others look for their support or company, which interferes with their capacity for emotional resonance. A humble person, on the other hand, is naturally inclined to serve. While narcissists are usually fantasizing about success, recognition, power, and money, a person who has cultivated humility tends to have a noble and generous heart.

When Freud discussed narcissism, he even suggested that the narcissist's libidinal energy is so invested in self that they're incapable of developing transference with a therapist (a key aspect of psychoanalysis), and thus, their therapy would fail. In other words, he believed narcissism has no cure.

## The narcissist's profile?

The DSM-5 can be consulted for a complete list of traits a clinician takes into account in order to reach a diagnosis of a narcissistic personality disorder. But without going into diagnoses, narcissists are those we commonly call "impossible people." They feel unique and special and do everything possible to infiltrate exclusive social circles. They seek and demand recognition, admiration, and respect, but it's difficult for them to make real friends. They have a sense of ownership over their partners and their children. Their possessiveness leads to jealousy.

People who have not developed a healthy self-esteem may fall for the charms of (covert) narcissists who may use their charisma to prey on what they perceive as vulnerable targets. They might offer to solve their problems, lend them money, listen to them, offer free advice, mentor, and even get them out of financial trouble. However, since they're exploitative, their behavior tends to seek personal gain. Too often, affiliations between narcissists and people who see themselves as "losers" evolve into codependent and abusive relationships. First applied to relationships where a partner unconsciously enables the other person's addiction, the term codependent was extended to describe relationships in which one person compulsively plays the role of a rescuer, thinking they can make others happy.

Because narcissists' self-value often depends on the appreciation from those they take care of, they may adopt the role of redeemers. However, if by any chance a narcissist's partner begins to grow as a person, becomes more independent and secure, and stops needing their protection and support, the relationship can go sour, and abuse can be accentuated. To restore their messianic role, narcissists often resort to gaslighting to create self-doubt and humiliate others, making sure they maintain their dependency. Criticism, judgement, and shaming become a habit. Verbal abuse and financial control may become extreme. Very often narcissism and abuse go hand in hand.

A study at the University at Buffalo School of Management revealed that men, on average, are more narcissistic than women.[123] They analyzed 475,000 surveys over the course of thirty-one years and found that men scored higher on narcissism at all ages. The same surveys revealed that men more often believe they're entitled to certain privileges and opportunities. The researchers concluded that the differences between the sexes resulted from sociocultural influences. They also found that narcissism was associated with other interpersonal dysfunctions, including difficulties maintaining long-term relationships, unethical behaviors, and a tendency toward aggression. We've already looked at how the environment—the type of society in which we grew up—shapes the way we adapt to different situations.

It seems symbolic that, as a humanity in the initial stages of the development of introspection (the ability to look inward), we've created the mechanisms and developed the compulsion to look at the world and relate to each other through all kinds of screens, large and small, as if they served as a substitute for the surface of the lake in which Narcissus saw himself and took to admiring his own beauty, falling in love with himself. As necessary as it is to develop self-esteem, it seems we are instead developing an unhealthy kind of self-love. Unsurprisingly, several recent studies find that we're becoming more egocentric and less empathic.

---

[123] M. Biddle, "Study: Men Tend to Be More Narcissistic Than Women." Release from University of Buffalo, (published online, March 4, 2015).

Scholars are still trying to sort out whether social media (Facebook, Instagram, etc.) contribute to increasing narcissism or if narcissists find in social media the ideal platform to behave in a way they wouldn't dare in real life. The discussion isn't settled. It seems to me that in a world where we've disconnected from our tribe, from mother Earth, from our emotions and instincts, we've developed a heightened need for others to see us, and the digital world offers us the perfect mirror/screen.

In this virtual world, we tell others, "Here I am (geographically). See my gastronomic preferences (photos of meals in restaurants, recipes). Read my favorite quotes (instead of our own musings, millions of quotes found on the internet or in a book, are published in social media); watch my videos. See my family, my dress, my cat, my grandson, my house, my new shoes. I want you to see who my friends are, my family. This is what I read. Let me show you what my political preferences are." All the while, waiting for a "like" or congratulations or condolences. It's like we're shouting, "Please reflect my image, tell me who I am, love me, validate me, please attend my events. Buy what I sell. Be as me. Like me."

It is true that socializing and undertaking actions together is a form of mutual support that provides a personal reward while dissipating physical and mental tension. It usually gives us satisfaction and security and increases our experience of belonging. But, dare I say this? Social media will never replace a face-to-face interaction between two or more human beings, because, among other reasons, the way this communication happens is faulty. Metacommunication is a key element of live communication. Made of nonverbal cues, like tone of voice, facial expressions, body language, metacommunication tells us how to interpret the message received. And it's absent from digital interactions. Neither the emojis nor the images we share can effectively replace it.

Western cultures have encouraged a sense of individual freedom that has swelled the boundaries around each one of us, but it also seems to boost narcissists' cravings for personal recognition and power. This contrasts with Eastern and Third World cultures, in which values such as humility, a sense of community, solidarity, empathy, and kinship between man and the environment still seem to be valued.

This discussion brings to mind a scene from the film *Seven Years in Tibet*, by director Jean-Jacques Annaud. It illustrates the encounter between two cultures during the Second World War: that of the protagonist, Heinrich Harrer, born in German Austria, and that of the Tibetan seamstress, Pema Lhaki. When Harrer and his compatriot Peter Auschnaiter both look for the seamstress' love, Harrer tries to dazzle her by displaying his own abilities and charms. To his surprise, Pema is seduced by Auschnaiter's simplicity, humility, attentiveness and sense of solidarity, not Harrer's pride or display of skills. She explains why: "This is another great difference between our civilization and yours. You admire the man who pushes his way to the top in any walk of life, while we admire the man who abandons his ego."

When Harrer, who had the rare privilege of being admitted as a refugee in Lhasa, the Sacred City of Tibet, becomes a tutor for the young Dalai Lama, he eventually ends up learning from the compassionate acts of his pupil. The young Lama represents a people with a strong pacifist tradition that contrasts with the aggressiveness of the forces that fought over the world during the Second World War.

There is a correlation between individualistic, competitive cultures, our level of self-esteem, and our ability to connect with others and be empathetic. Take into consideration that interpersonal bonds are formed and maintained thanks to our capacity for empathy.

A group of scientists from Michigan State University interviewed 104,365 adults in sixty-three countries to rank their ability to react and connect with others, both emotionally and intellectually. The study[124] was led by a psychology professor, William Chopik, who's trying to understand why the United States ranked seventh (Ecuador was first) and why the predominant culture in the US has become more and more individualistic. Chopik noted that the mental state of Americans has changed in recent decades, focusing more on the individual and less on others, which makes it difficult to establish meaningful and close relationships.

In the study, the questions sought to measure empathy by determining the respondent's ability to imagine another person's point of view. The five most empathic countries were Ecuador, Saudi Arabia, Peru, Denmark, and the United Arab Emirates (in that order).

Another study led by Sarah Konrath, from the Institute of Social Research of the same university, compared surveys of fourteen thousand American university students conducted between 1979 and 2009. Their study suggests that "college students today are less likely to 'get' the emotions of others than their counterparts 20 and 30 years ago. Specifically, today's students scored 40 percent lower on a measure of empathy than their elders did."[125] No wonder they're considered the most narcissistic, competitive, and individualistic generation in recent

---

[124] W.J. Chopik, O'Brien, E., Konrath, S.H. "Differences in Empathic Concern and Perspective Taking Across 63 Countries." *Journal of Cross-Cultural Psychology,* 48(1), 23-38, (Oct. 2016) doi.org/10.1177/0022022116673910

[125] J. Bryner, "Today College Students Lack Empathy," *Live Science*, (May 28, 2010).

history, according to Konrath, who believes that this phenomenon is related to the number of hours young people spend absorbed by violent video games, which could make them insensitive to someone else's pain.

Jeremy Rifkin, on the contrary, points out that, in a globalized world, diverse people cross paths more frequently, and this has forced us to develop an empathic sensitivity and to expand our consciousness.

Narcissism and lack of empathy measured by the surveys mentioned above seem to rise as a reaction to a demanding and perfectionist culture, to which Rifkin also alludes in his book. While pushing for adherence to the highest standards of moral perfection, this type of culture also generates shame, envy, jealousy, and contempt for oneself and others.

Surely, it's also true that globalization has given rise to an increased in-group empathy bias that might explain the explosion of protectionism and nationalism, as well as racist and discriminatory expressions in general.

But Rifkin is an optimist. He sees that "a younger generation is fast extending its empathic embrace beyond religious affiliations and national identification to include the whole of humanity and the vast project of life that envelops the Earth."

## Fearing you're not enough

*And above all things, never think that you're not good enough yourself. A man should never think that. My belief is that in life people will take you at your own reckoning.*

—Isaac Asimov

I'm saddened by how often I see patients who come to my office overwhelmed by the feeling that they're not enough (in their jobs, for their partners or their parents, for their children). No matter what they do, how dedicated they are to their work, how much they've grown emotionally or intellectually, they tend to feel little and insufficient, and they're unforgiving of their own mistakes. This feeling of not being enough is a source of suffering and sometimes despair.

So many people grew up believing that love has to be earned and that others can't love them just for who they are. This feeling comes from the way you were raised in a competitive society, previous relationships (with parents, siblings, teachers, friends, partners) and is framed by a competitive culture in which the goals of becoming famous, rich, and powerful are the driving forces of our lives. Then, if you aren't a millionaire hitting the headlines of newspapers or TV (only 6 percent of people in the US are millionaires), you're a failure. More often it's about a more modest goal, such as making a six-digit salary and owning a four-bedroom house. In any case, if you haven't achieved these goals by a certain age, you tend to see yourself as a disappointment.

The American dream is no longer about freedom, democracy, civil rights, and the opportunity to climb the social ladder with hard work. If many people used to come to the United States with the dream of providing their children with an education to guarantee a future different from that of financial hardship, now more parents seem to dream of their children becoming billionaire financiers if not famous soccer players or glamourous models. But, realistically, how many can reach those goals?

There aren't enough opportunities (I'm not referring to more shows fueling a dream of leaping to fame in a stroke of luck) for those who, since childhood, have demonstrated special talents for dancing, singing, comedy, art, mathematics, or science. More than once I've felt saddened by *Got Talent* contestants' stories of sacrifice and broken dreams. So many people have been ignored by the world until they showed up on television.

Consider Paul Potts, forty-seven, the son of a bus driver and a supermarket cashier, who won the contest *Britain Has Talent* in 2007, and Susan Boyle, fifty-six, who reached second place in the 2009 contest. At his first audition, the image of Potts[126] facing the judges provoked mixed reactions from the audience. Among the reasons motivating him to compete he said, were his low self-esteem and lack of self-confidence. However, doing what he loves (singing opera), makes him "feel part of a whole so big that nothing else is missing." This Carphone Warehouse manager only fully launched his career as a singer after his superb performances at a contest broadcast to millions of people. He has now traveled the world, and his music sells.

Susan's case is similar. When she was a child, she was bullied by her peers, many times for behaviors perceived as odd. Her parents believed she might have suffered brain damage. She dreamed of becoming an opera singer. At her first audition, the judges looked at her full of skepticism. She hadn't had a formal music education, and her appearance and behavior were a bit unusual. But within a few seconds of beginning to sing, there were tears in the judges' eyes. After becoming a finalist in the contest, Susan sung in front of the Queen of England and

---

[126] "Paul Potts stuns the Judges while singing 'Nessun Dorma.'" Watch on youtube.com, *Britain's Got Talent* channel.

for President Obama and has sold up to twenty-five million records. Her dream came true. Susan has been diagnosed with a high-functioning form of autism known as Asperger's syndrome, for which she could have received adequate support if she hadn't been misdiagnosed as a child.

A truly democratic society should stimulate the development of talents, skills, and abilities from childhood. Being able to develop those talents would not only contribute to the society as a whole but also significantly nourish a healthy self-concept. However, supporting the growth and development of a gifted child—including their intellectual, emotional, social, and physical domains—should not be left to private initiative or chance.

Exiting the trap of feeling we're not enough requires significant changes in the structure of society, a change in priorities, a change in how we define progress, which is currently based on, consumerism, competition, and individualism. Liberation comes from a sweeping change in consciousness of who we really are, how we depend on each other, and what difficulties we need to overcome to build a just society. We don't conquer love, we consciously build it, daily.

It's true that life challenges make us stronger, and difficulties are inescapable. Our goals are a beacon on the horizon; we follow the light so as not to get lost. However, perfectionism accompanied by panic over making mistakes, perfectionism that leads to feelings of inadequacy, constitutes an obstacle. Pursuing impossible ideals leads us to be defeated by reality. Seeking excellence instead (the best we can do or give of ourselves) includes risks while excluding the fear of failing.

In *The Danish Way of Parenting: What the Happiest People in the World Know About Raising Confident, Capable Kids*, (Penguin, 2016) Jessica Joelle Alexander and Iben Sandahl share with us the secret for a happy society. Danish people raise their kids to be empathic, resilient, solidary, and joyful, most of all by letting them play instead of imposing impossible goals since birth. Danes focus on developing children's internal locus of control, a sense that what you achieve results from your own efforts and abilities and not someone else's demands. Sandahl says, "We try to provide them with a scaffolding for their development to let their self-esteem climb. In the long run this makes happier children who grow up to be happy adults – and then the cycle repeats itself."

Cultivating qualities such as flexibility, acceptance (of ourselves and others), authenticity, and introspection (which allows us to take responsibility for our actions), without neglecting our duty to others and the planet, that is in itself excellence. The search for perfection, on the other hand, results in feeling that any achievement below a certain (often impossible) ideal is unacceptable. For a perfectionist who lacks self-love, errors become failures, not opportunities to improve. It's a very different story to be brave enough to embrace the challenges that help us grow, be productive, and serve others than being driven by a desire to please, fit in, be admired, or achieve fame.

All life paths, all kinds of work, have an implicit dignity that comes from our social contribution. All work includes a certain degree of service to the community, even if it doesn't seem obvious to us. We depend on the people farming the food we consume, the driver that transports it, the employee who stores and classifies it. The very tissue of society is made of the way we complement and support each other. We all contribute to the social fabric, we're all valuable.

# Love for your partner

*But let there be spaces in your togetherness,*
*And let the winds of the heavens dance between you.*
*Love one another, but make not a bond of love:*
*Let it rather be a moving sea between the shores of your souls.*
—Khalil Gibran

## In my practice: couples learning to develop empathy

*The couple sitting across from me are trying to save their marriage. When I listen to them talking about their disagreements, I wonder how or why those vows of eternal love sworn on their wedding day no longer guide their actions. While they talk, my mind builds a different version of the story their words tell me. I'm moved by their despair, their impatience, the anguish in their voices. I see their bodies tensing. He throws his body against the back of the sofa when he hears her complaining about the way he treats her and how she wants to leave him. She shuts up and leans forward, crossing her arms across her torso, her hands reaching across and massaging her arms, soothing herself. Or getting ready to protect herself from a blow that might never come but is always a threat. Maybe she's afraid of expressing all of her truth—that loud secret he hasn't wanted to hear in so many years of being together. She has patiently endured his criticism, his reproaches, his shouting. She learned from her mother that "worthy women suffer in silence." Society expects her to be patient, understand her husband, fight for her marriage, be up to the until-death-do-us-part promise. He feels he's going to lose her no matter what, and her words are certainly not giving him hope. He doesn't seem to have insight into what his share in their situation is, and he keeps accusing and blaming her. With each one of his words, she slips further away.*

But—here they are. If they've come to seek the help of a therapist, as a last resort perhaps, this may indicate that there's still something to rescue in this marriage that, apparently, is sinking. It doesn't matter if his name is Pedro or James. It doesn't matter if her name is María, Juana, or Alexandra. It does not matter whether they come from this or that background or what their professions are—doctors, bankers, or blue-collar workers. The story is basically the same. On the surface, their disagreements seem to be about something simple, like how to run the house or time management, but the real issue is their particular way of understanding relationships, their subconscious definition of what love is and their particular understanding of how it needs to be expressed.

He complains about her lateness; he doesn't feel she's taking good care of him. She complains that he is extremely controlling. She wants him to come straight home from work to participate in the household chores. She wants him to understand how the children need his presence. He demands absolute loyalty, that she support him in everything, that she recognizes his authority and understand how he's breaking his back to make a living for the family.

He points a finger at her and tells her she doesn't give him enough love. She says he doesn't understand her need for personal space or acknowledge her contributions. She explains that when she doesn't want sex, it's because she's exhausted or maybe because she needs to be seduced into the great encounter with a little bit of romance, a flower, a card, a loving call. For him, she says, nothing she does is ever enough, and his jealousy has killed sexual desire. And so on.

We establish ground rules to start therapy:

The office represents a safe space. Whatever is spoken here, stays here. No accusations will be made. They will speak in the first person, taking responsibility for their part in what's happening. If they've learned to respond to tensions or disagreement—a perceived threat to the relationship—through fight-or-flight, they could learn new responses now. Hopefully, they will become more present and learn to respond empathically.

The sessions are confidential, but we establish a no-secrets policy. It's understandable that each individual wants to be validated, to feel he or she is right. However, the therapist will not take sides. If one of them calls to complain about the other, that will be discussed in the sessions.

If a therapist took sides, that would stimulate rivalry, the second response to stress. Competing is already killing the relationship. It has become more important to be right than to rescue their relationship. Key question is, can they put themselves into each other's shoes?

Therapy's purpose is to save the relationship, if possible—to protect the family. And if they didn't have a happy marriage, at least they can work on achieving a smooth divorce process, preventing the deepening of the wounds, protecting the children.

Focus will be on strengths and not on weaknesses. To activate the third response to stress, they need to cultivate appreciation and facilitate empathy. Pointing at the other's shortcomings, criticizing, focusing on what the relationship is not or should be, deepens the gap and awakens the fight-or-flight response. A home must create conditions for empathy, cooperation, and mutual support.

While feelings are being processed, they will speak in the first person to express their feelings, their personal outlook on what happened. Too often people find relief in blaming each other but this only deepens the wounds.

One of the goals in therapy is to learn to take responsibility for one's own actions and choices. When an individual takes responsibility for their part in the relationship, they can stop feeling and behaving like a victim. The moment a person is in charge of her or his life, they start seeking solutions. If severe verbal, emotional, or physical abuse is present, the best resolution is often to leave the relationship, not to escape from the other but to embrace the possibility of starting a new life.

Very early in my career as a family therapist, I realized that after the initial evaluation of a couple and the agreement on a plan for treatment and interventions, it's key to educate them about ways to love each other. Up to the moment their conflicts overwhelm them, they have just been repeating relational patterns that they uncritically learned from their family of origin or from other models that society has presented to them. An alternative way of relating needs to be explored. I invite them to examine their biases.

To strengthen their capacity for empathy, clients are encouraged to disclose their feelings (not to focus on ideas) while practicing active listening, which involves paying attention, avoiding judgment, reflecting on what's being said, asking for clarification, summarizing what was understood, and then sharing feelings elicited by a shared event and the different ways they recall it. When they look into each other eyes, they remember the love they felt for each other, and often the spark is rekindled.

Researchers have found a link[127] between gazing into each other's eyes and affection, love, and lust.

When I see couples, I ask them to sit facing each other, looking into each other eyes, while they take turns listening and talking. They're instructed to echo what the other is saying while, at the same time, exploring their inner reaction to the words being said and trying to respond to the other person's feelings in a loving way. I later explain to them that emotions are contagious and therefore emotional sharing and empathic concern usually follows. With luck, after some soul searching, people learn to look at things from the perspective of the other. [128]

However, there's very little hope of saving a relationship when empathic inference is missing or when people find it difficult to communicate feelings or are not open to acquiring these skills. An egocentric view becomes an impediment. Nevertheless, there's no doubt that unless a person is a narcissist or a sociopath, they can develop a capacity to experience empathy and love.

But how do you teach others to love? Is there a universal way of loving? And if what a counselor or an educator has learned about love comes from her growing up in a complex family, from her own failures in relationships, and from her own parenting shortcomings, would she be competent enough to teach others how to love? Will it be enough, pertinent, or even valid to transmit her own experience to others? Therapists are certainly trained to listen more and advise less. It's disquieting to wonder if the therapist's own discoveries (hopefully explored with her own therapist), the result of her personal journey, would apply to other people who have not yet processed their own issues. A danger I see in advising others is that they may end up following advice that's not relevant.

Self-help books or even experience will never be enough to learn all that's needed about this topic. I advocate for therapists to explore their traumas, subconscious biases, and limiting beliefs, so that their baggage doesn't turn into countertransference and interfere with their service.

I tend to see my office more like a laboratory where clients and I come together to explore new paths and healing solutions. Clients come to a place where we all learn, if we approach the adventure of therapy with open minds. I invite clients to find out what they understand *love* to mean.

---

[127] J. Kellerman, J. Lewis, and J. D. Laird, "Looking and Loving: The Effects of Mutual Gaze on Feelings of Romantic Love," *Journal of Research in Personality 23*(2) (1989): 145–161.

[128] J. Schulz, "*Emotions Are Contagious: Learn What Science and Research Has to Say about It,*" Michigan State University Extension (August 16, 2017).

What's valid for the couple in the relationship?

What works for them or their family?

We're intent on breaking limiting beliefs and semi-truths that were embraced subconsciously as unquestionable and that prevent us from letting down our guard. I dare them to talk about their deepest secrets with an open heart, without judgment. In many cases they're acting out of a great despair, born from seeing their relationship going sour and not being able to correct course. There's nothing more painful than losing the emotional investment made in the relationship, not being able to trust the other any longer, or feeling that the illusion of a forever has been broken.

In the office, we examine issues in which we're all still quite illiterate: gender issues, the changing roles of modern women, the definition of a violent relationship, etc. But a topic that couples often avoid discussing openly in front of a therapist is their sexual life, which somehow gets detached from the context of the problem that afflicts them. Good sex also requires empathy (feeling that the partner deeply understands what you feel, what you want). Since I'm not a sex therapist, patients who come to me sometimes don't feel they've come to discuss their sexual life. But sooner or later, it needs to be addressed, and not surprisingly, in most cases it's found at the core of their conflicts.

Belgian psychoanalyst Esther Perel, who has promoted the concept of "erotic intelligence," believes that social taboos and ideals of domestic equality have compromised the healthy expression of eroticism in modern relationships. She says that it's a shame that sex is not addressed more directly and thoroughly.

Perel said,

> The body often contains emotional truths that words can too easily gloss over. The very dynamics that are a source of conflict in a relationship—particularly those pertaining to power, control, dependency, and vulnerability—often become desirable when experienced through the body and eroticized. Sex becomes both a way to illuminate conflicts and confusion around intimacy and desire and a way to begin to heal these destructive splits. Each partner's body, imprinted as it is with the individual's history and the culture's admonitions, becomes a text to be read by all of us together.[129]

[129] E. Perel, *Mating in Captivity: Unlocking Erotic Intelligence* (New York: HarperCollins, 2009).

## Amo, ergo sum

From the *Kama Sutra* to the love poems of Sappho of Lesbos, romance writings populate ancient literatures all over the world, although Western anthropologists and social historians have largely ignored them.

Perhaps influenced by some contemporary historians and philosophers, I used to believe that romantic love was merely a bourgeois invention, perhaps the legacy of the type of romance that characterized European medieval culture. I recognized the existence of attachments, loyalty, desire, and the need for a companion. Of course, in many cases (even in modern times), I was aware that certain unions were driven by economic interest or a search for a status obtained through a convenient marriage. But over time, my cynicism has given way to optimism, and despite the pervasive high rate of separations and divorces (significantly higher in so-called developed countries), I now tend to believe that because love is our essence, lasting romantic love is definitely possible.

Recognizing romantic love as valid comes from, among other things, realizing that the choice of a lifetime partner is often subversive. That is, the choices people make don't necessarily respond to what society's norms or the family would deem a good, convenient, match. Nevertheless, there is no doubt that commercial interest in a consumerist society has added unnecessary frills to love, such as the celebration of special (good-for-business) days like Valentine's Day. And weddings alone have become a seventy-billion-dollar business in the United States.

In our discussion about the neural circuits of love, we said that a couple's love comprises three elements: sexual attraction, romantic love, and attachment. Traditionally, anthropologists had assumed that where life conditions were precarious, romantic love didn't have many possibilities. Moreover, some assumed that romantic love was exclusive to Europe and North America. But now we have evidence indicating that these elements might be common to all cultures.

Doctors William R. Jankowiak and Edward F. Fischer[130] published the first intercultural study systematically comparing romantic love in various cultures. Their conclusions were based on a survey that covered 166 cultural groups, and they found clear evidence that romantic love was

---

[130] W. R. Jankowiak and E. F. Fischer, "A Cross-Cultural Perspective on Romantic Love," *Ethnology* Vol. 31, No. 2, (April 1992): 149–155.

known in 147 of them (89 percent). In the remaining nineteen cultures, according to Dr. Jankowiak, the absence of conclusive evidence was probably due to inadequate monitoring by anthropologists rather than a lack of romanticism in the group. The above research was mostly based on reports from volunteer informants.

One of the testimonies they recorded is that of Nisa, a Kung woman among the Kalahari Bushmen, who made a clear distinction between the affection she felt for her husband (a relationship she described as "rich, warm, and safe") and what she felt for her extramarital affairs (which she described as "passionate and exciting," although fleeting). Of these affairs, she said, "When two people are first together, their hearts are on fire, and their passion is very great. After a while, the fire cools and that's how it stays."[131]

In *A Natural History of Mating, Marriage, and Why We Stray*, (W. W. Norton, 2016), Doctor Helen Fisher says that romantic love evolved around two million years ago with *Homo erectus*, when the collaboration of a partner became essential in obtaining food for the offspring. The biochemistry and neural circuits related to the feeling of love would have evolved since then.

Although it's now commonly accepted that the experience of romantic love is universal, there are significant differences between the most prevalent emotional expressions in each culture, which can be understood as the result of particular styles in the regulation of emotions and also the influences of parenting and traditions.

In the West, the individual has precedence over the community. Personal choice is important. Arranged marriages are now uncommon.

Mendy Wang, at the Feng Chia University in Taiwan,[132] says: "[Eastern cultures] tend to be submissive than to be aggressive [*sic*]. They think that being a leader is difficult, and they don't like to take heavy responsibilities. Also, they are contemplative. They think more and do less; on the contrary, Western people do more and think less. They are diligent and assertive; they love to be leaders and make decisions."

Wang says that while in most East Asian countries, which have stronger patriarchal traditions and a strong sense of duty, "good couples" are interdependent and adjust over time to their mutual expectations, in Western countries there seems to be more conflict, and expressions of anger are normal.

---

[131] Ibidem.

[132] M. Wang. "Eastern vs. Western Culture." *FLL Mosaic*, June 2017, Feng Chia University, Taiwan.

In the East, based on a sense of duty, in order to live up to the expectations of the other, individuals must be aware of themselves and grow as people. This seems to contribute to building stable relationships to the extent that individuals align with established social norms. Honesty is an important value in these cultures. Anger, on the other hand, is not accepted because it disturbs harmony. The above characteristics mark the type of relationship formed and the predominant emotions.

## Choosing a partner?

*What many people call loving consists in choosing a woman and marrying her. Choose it, I swear, I've seen it. As if you could choose in love, as if it were not a lightning that breaks your bones and leaves you stuck in the middle of the yard.*

—Julio Cortázar

The degree of social development of a country has been correlated with the tendency to value or not value certain practical and material aspects when choosing a partner. In collectivist and less economically developed countries, social status and good health seem to have the most value. In these countries, casual and uncompromising or playful love also seems to be valued more, but marriage is still seen as a sacred institution.

The process of globalization and the evolution of socioeconomic status and social structure have brought important changes to relationships, especially noticeable in a lesser differentiation of gender roles. With a higher income and economic independence, individualism has been reinforced, and equal opportunities for men and women have led to an increase in the importance assigned to subjective feelings and personal decisions regarding how a relationship is framed.[133] However, a higher independence also seems to correlate with the increase in divorces and the decrease in birth rates.

In addition to the role that physical appearance has in the attraction we feel for others, psychological traits processed unconsciously are predictive of an emotional connection and a long-term relationship. Please take into account that the choice of a partner is usually influenced by a built-in ideal, an unconscious profile—a profile based on the

[133] F. Bejanyan, T. C. Marchall, and N. Ferenczi, "Romantic Ideals, Mate Preferences, and Anticipation of Future Difficulties in Marital Life: A Comparative Study of Young Adults in India and America," *Frontiers in Psychology* (December 2, 2014).

relationship between our parents, our past relationships with parents, our childhood experiences within the extended family, and norms and models offered by society.

Individuals with low self-esteem might find themselves attracted to interact with people similar to those who've already humiliated them, who've made them feel inadequate and not enough. In the same way, unconscious reasons drive people to feel attracted by those offering a conditional type of affection. Psychoanalysis has explained it as an unconscious psychological attempt to rewrite and heal personal history.

In the *New Introductory Lessons on Psychoanalysis* (Carlton House, 1933), Freud said, "There are people in whose lives the same reactions are perpetually being repeated uncorrected, to their own detriment, or others who seem to be pursued by a relentless fate, though closer investigation teaches us that they are unwittingly bringing this fate on themselves."

For those with very busy lives or those who are too shy, the internet offers the option of finding a partner in a more rational way. According to the Pew Research Center (statistics from 2016), "11% of American adults—and 38% of those who are currently 'single and looking' for a partner—have used online dating sites or mobile dating apps."

In a digital era, online dating is becoming increasingly popular with young people. It's so easy to just browse and find an online dating site. Fill out a form listing your own attributes and those of a preferred partner. Post a photo, and, just as if it you were buying any merchandise, get a list of comparable and possibly compatible candidates. This might be a less romantic but practical way of establishing a relationship, supported by studies concluding that people who have matched this way have a lower risk of divorce.

John Cacciopo[134] and colleagues addressed this issue with a national sample of 19,131 respondents who married between 2005 and 2012. They reported that: "More than one-third of marriages in America now begin on-line. In addition, marriages that began on-line, when compared with those that began through traditional off-line venues, were slightly less likely to result in a marital break-up (separation or divorce) and were associated with slightly higher marital satisfaction among those respondents who remained married."

In *Why Love Hurts: A Sociological Explanation* (Polity Press, 2012), the sociologist and anthropologist Eva Illouz, expresses her

---

[134] J. T. Cacioppo, S. Cacioppo, G. C. Gonzaga, E. L. Ogburn, and T. J. VanderWeele, "Marital Satisfaction and Break-Ups Differ across On-Line and Off-Line Meeting Venues," *PNAS* 110 (25) (June 18, 2013).

disagreement with online mating, "The rationalism which characterizes late modernity has led to uncertainty and irony," she said, expressing that we shouldn't behave as consumers in the matters of the heart. Choosing a partner online, as well as any attempt to choose rationally destroys the erotic and can inhibit a long-term commitment. For her, love has become "the object of endless investigation, self-knowledge and self-scrutiny."

Polish sociologist Zygmut Bauman was also critical of the process. He said in an interview for The Conversation that in online matching, "The underlying idea is that an object of love can be assembled from a number of measurable physical and social characteristics. In the process, the most decisive factor gets forgotten: the human person.[135]"

Online matching might miss a key element of instinctual attraction that happens in a personal encounter. The anthropologist Helen Fisher studied hundreds of couples in love and found that the human body knows in the first seconds of meeting someone whether or not there is attraction for the other person. The mind is that quick to make a judgment. But, often, we're simply attracted to what's familiar to us.

In the sixties, Robert Zajonc, a social psychologist at the University of Michigan, conducted several studies in which he demonstrated the existence of a phenomenon known as "mere exposure effect," also known as "the principle of familiarity," which states that "the mere repeated exposure of an individual to a stimulus object enhances his or her attitude toward it." [136] In other words, we almost immediately develop a preference for what is known to us without a cognitive process mediating it. This seems to be especially true for people with high needs for relational closeness.

Love relationships begin with a load of expectations. In an egocentric and utilitarian society, the couple is viewed from the perspective of the self. The question often is: "What do I get out of this relationship?"

Depending on the socioeconomic status and the need for financial stability, a good provider is often sought. For emotional stability, people tend to seek a balanced partner, able to modulate the other person's emotions. To raise a family, people often seeks someone capable of being a good parent. To compensate for our weaknesses, we might seek

---

[135] Karenarchey, "Zymut Bauman on Love and the Internet." January, 2017. Retrieved from https://conversations.e-flux.com/

[136] R. B. Zajonc, "The Attitudinal Effects of Mere Exposure," *Journal of Personality and Social Psychology*. Monograph Supplement Volume 9, No.2, Part 2 (1968). Pdf retrieved online June 25, 2019 at https://pdfs.semanticscholar.org.

a strong partner. Ideally, we should be looking for a person who's capable of empathy and love.

The expectation that another person will make you happy, the belief that being happy is the purpose of any relationship, becomes a perfect ground for the establishment of a codependent relationship. Nobody can make us happy. We can't make others happy. But we can find joy in deep connections with others.

## We contribute our inner world

Psychotherapy and psychoneurobiological education could help people understand the subconscious motivation for their choices and failed attempts at finding the ideal partner. It could help not only by exploring and processing the unresolved business from the past, but by helping people believe in the possibility of rewiring their brains (yes, as we have already said, brains can change) in order to embark on the adventure of healing and breaking old patterns.

A person doesn't exist on an island. In addition to our personal experiences, each one of us is the result of the context in which we've existed. In that sense, you could say that every relationship is cross-cultural and every family constitutes a micro-culture. Although two individuals might share the same values and beliefs, and even a set of behaviors, each family has its own narrative, a series of shared experiences that have determined their vision of the world and the establishment of similar traits and relationship patterns. Our identity has been formed around not only our ethnic origin, mother tongue, and family or social beliefs, but also around socially accepted behaviors, modes of resolving conflicts, degree of support from the extended family, self-assessment, ideas of right and wrong, aspirations and interests, social status, religious ideas, and concepts of life and death. That is, when we enter into a relationship, whether we realize it or not, we do it with a particular vision of the world and a set of biases and prejudices.

For all the above reasons, we reproduce customs and habits from the family of origin or of any real or fictional family we established as a model. The family and social context in which each one grew, the emotional language that the family used or didn't know how to use, bears a specific weight. In short, we reproduce the relational patterns that were learned at home or in school. Other influences come from the literature we read, the movies we watched, the models we were exposed to, which lead us to create those ideal partners in our minds and bring certain expectations to the relationship about what the other should be and how they should behave.

Couples' therapist John Gottman, author of *The Seven Principles for Making Marriage Work* (Harmony Books, 2000), has researched, written, and taught classes on how to predict if a marriage has a future and how to work for it. Based on the systematic observation of interactions between couples, Gottman drew conclusions about what obstacles interfere with harmony in a relationship, and he made recommendations on how to establish and strengthen intimacy and know each other more deeply. When the author speaks of intimacy, he focuses as much on the erotic aspect of the relationship as on the empathy and compassion the couple could experience. To encourage them, he designed "love maps," a series of questionnaires and games that fill in the gaps of information about how we define love, and what we've learn consciously or unconsciously about how to love. Assumptions built from former personal experiences fill these gaps. They feed the individual's doubts and fears, usually generating misunderstandings and conflict. We quickly turn defensive when the other person points at some personal aspect we consider flawed.

If we seek to be in a relationship to satisfy emotional needs—such as a need for approval, company, or acceptance—our wants will be greater than the love we could offer. In our search, we would be regressing to that first stage of life, in which love is mostly egocentric, where I love you because I need you—possessively. This is the kind of love in which I want to become part of you to complete myself. Or, I would try to control you and mold you in order to satisfy my needs.

Another fallacy is to love another because he or she is just like you. That would be the equivalent to loving oneself in the other (a form of narcissistic love).

## Is love blind?

> *JESSICA: Here, catch this casket; it is worth the pains.*
> *I am glad 'tis night, you do not look on me,*
> *For I am much ashamed of my exchange:*
> *But love is blind and lovers cannot see*
> *The pretty follies that themselves commit;*
> *For if they could, Cupid himself would blush*
> *To see me thus transformed to a boy.*
> —Shakespeare

Let's critique socially transmitted assertions, such as that tough love is a good way to show you care, that you need to enforce particular constraints on your partner (or your children) to make them take responsibility for their actions.

Responding to male domination and any form of domestic violence, we've adopted the motto that real love doesn't hurt. But we also need to understand that the risk of pain is inherent to relationships. Yes, broken hearts do happen. Yes, there is some suffering from being separated from the ones we love and there is pain when our relationship is not going well. Actually, a neural association has been found between social and physical pain and if you're heartbroken, the pain is so real that you can relieve it by taking a painkiller. The feel-good hormones are over and you go into withdrawal.

But "tough love" is a, different, complicated matter. If we mean setting clear boundaries, establishing limits, clarifying the rules and norms, these help families function healthily. But sometimes the term is misunderstood to mean "punishing" the other by withholding affection, making the other jealous, stonewalling, criticizing, showing contempt, and using sarcasm. And these are not at all expressions of love.

Abandonment, betrayal, or abuse hurt people. And if a relationship involves a toxic form of love, you probably need to run away from it.

It's not difficult to figure out the origin of the belief that suffering is intrinsic to love. We've too often been comforted with a sigh, a shrug, and a resignation of rights and dignity leading a victim of abuse to accept their situation. It goes with the statement that love is blind. As if there was nothing to do about it, as if the destiny of a victim is unescapable when love is (supposedly) involved.

True, mothers and fathers suffer for their children: their grievances, their disappointments, their struggles hurt us. Friends join us in our anger and love disillusionments, and they cry with us. The loss of a loved one can be extremely painful, traumatic. Magnetic resonance imaging shows that brain regions that get activated when a person has physical pain overlap with those areas activated when a person experiences the anguish of a breakup. A condition known precisely as "broken heart syndrome," characterized by chest pain, palpitations, and inflammation, has been described.

In her book, *Heartbreak: New Approaches to Healing—Recovering from Lost Love and Mourning* (Salem Author Services, 2011), Dr. Ginette Paris, professor of Jungian and archetypal psychology at Pacifica Graduate Institute, in Santa Barbara, California, explains that breaking up a relationship is a traumatic event that requires appropriate treatment. The book combines psychology with neuroscience to explain what we feel when we experience losses.

Another painful aspect in a relationship that deserves mention is jealousy, which leads to frequent disputes, tensions, and aggressions. In some cultures, jealousy is even considered a symptom of a greater love.

Some psychoanalysts after Freud have hypothesized that jealousy originates in what is known as a "narcissistic wound," a background of shaming, abandonment, and betrayals experienced directly or indirectly by a subject, making them especially vulnerable and reluctant to reveal their true self. Others see jealousy as the result of projection (I know I'm capable of betraying you; therefore, I watch you, so you won't betray me).[137]

More recently, cross-cultural studies have shown that jealousy is typical of cultures that tend to be possessive, assign great importance to marriage, and believe sexual gratification should happen exclusively within it.[138] But no, jealousy is not an expression of love, it is related to a perception of the other as property, and it can be deadly.

Up until relatively recently in Colombia, in cases of femicide or violence against women, judges handed down lighter sentences if the defendant (usually a male) could provide proof that he had acted in a "state of anger and intense grief" caused by the adultery of his partner. Although the criminal code has been reformed, the laws continue to be lenient on a perpetrator claiming to have committed the crime under the influence of those emotions. And we still read news stories about the practice of stoning in Middle Eastern countries.

The myth that a man closely surveils the wife because he loves her so much is just that, a myth.

Evolutionists and sociologists theorize that jealousy is related to the evolution of the family. Once private property appeared, group marriages, or syndyasmian marriages ended. Syndyasmian marriages were a typical form of organization of the family that existed long ago between the stages called savagery and barbarism. One man married one woman, but men were not bound by a commitment to be faithful. With private property came monogamy. The purpose of monogamy was undisputed paternity. Now that some men possessed tools and land, they

---

[137] M. A. Neal, E. P. Lemay, "The Wandering Eye Perceives More Threats: Projection of Attraction to Alternative Partners Predicts Anger and Negative Behavior in Romantic Relationships," *Sage Journals* (October 10, 2017).

[138] D. Sikelianou, G. Georgakopoulou, and I. Pandiri, "Cross-Cultural Differences and Similarities in the Expression of Jealousy in a Couple: A Comparative Study Between Couples of Greek and Albanian Descent," *International Journal of Psychosocial Rehabilitation* Vol 19(1) (2015): 24-56.

wanted to know who their heirs were. Before private property, there was nothing for the children to inherit. This would also provide an explanation of men's sexual jealousy of their wife's encounters with other men (risk of pregnancy). Emotional jealousy was more typical of women, who feared that if their man became a provider for a new partner it would be to the detriment of her progeny.

A multidimensional vision would explain love and jealousy in terms of evolutionary, genetic, cultural, economic, psychological, and physiological facets.

Patriarchal ideology—asserting male privilege—has justified violence against women since medieval times and this ideology is still transmitted through expressions justifying jealousy or mistreatment of the partner. There is for example, a common saying in Spanish, "*porque te quiero te aporreo*." An approximate translation in English is, "I'm tough with you because I care for you." But just the opposite should be true, that because I love you, I should do everything possible to avoid doing or saying anything that hurts.

## Don Juan reruns

What is the common storyline in soap operas? Who does the female protagonist fall in love with? The bad guy, the difficult one, the one who betrays her, the impossible, the unfaithful, the forbidden. In the meantime, the loyal friend, the confidant who supports her at all times, becomes the fool who falls in love with her, even though she only looks for him in times of great despair. And how should we understand the fact that her fine friend falls for a woman who looks for him only when it suits her, humiliating him if she so pleases? I don't recall any soap opera ending with the protagonist and her loyal and silent admirer being together.

Those stories are based on a truth that any psychologist knows well: what is forbidden, what is difficult, fuels passion. They often portray the Don Juan hero-villain archetype: that guy who lives for romantic pursuit and seduction, and preys on women for the sake of the conquest alone. The stories are based on the pretense that when Don Juan finds "the one," a unique woman, she will show him the way to love and redeem him. This is, among other things, a model for codependent relationships.

Hollywood makes billions thanks to epic love stories that follow but variations of a fairly standard script: Joe meets Jill. Difficult circumstances hinder their love. At some point the truth is revealed, obstacles are overcome, or the protagonist is transformed, and Joe and Jill can finally be together and live happily ever after.

We must ponder what influence these stories have on our vision of relationships, especially in a society that doesn't teach us the language of love, train us to resolve conflicts in nonviolent ways, or help us develop empathy?

In the story we tell ourselves, conquering the love of the Casanova who despised us or abandoned us would somehow prove that we're worthy of being loved. On the other hand, the one who shows us unconditional love is pretty much an idiot. He's still there for us after we disrespect and ignore him. Why would we love such fool?

Let's make sure we're clear on the fact that the typical passionate love of the first stage of a relationship is different from the compassionate or empathic love that results from strengthening the bond. How many marriages dissolve just after the passion and the mystery of the first few years evaporate? But also, how many reap the fruits of maybe a less exciting time, where people grow together and commit to building a deep partnership in the day to day? A camaraderie that requires the ability and willingness to listen, take responsibility for one's own actions, express feelings openly, experience empathy and compassion, and forgive.

## Myths of love sung

Music both reflects and contributes to the generation of our culture, influencing our beliefs and behaviors. It transmits messages that perpetuate the status quo, even though lyrics often express social dissent.

There is often an eroticization of gender violence in music. It correlates with patriarchal models in which men still have a higher status and invite women to be sexually "liberated" (or, rather, sexually available).

On a positive note, women have also used music to protest against the corruption of those at the top, affirm their rights, and formulate powerful statements about their strength.

I'm not an expert on hip-hop or rap, but I've stumbled upon the interesting lyrics of Ani DiFranco and Queen Latifah. And, of course, of Little Mix with the martial song "Salute," with a call to ladies to awake.

We must also honor Aretha Franklin's voice and her popular song from the sixties where (though timidly) she demands "just a little bit" of R-E-S-P-E-C-T.

For ages, songs have reflected (and supported) the patriarchal status quo and its relationship patterns. Even the romantic bolero and the ballad have been guilty contributors to a culture of male privilege, spreading illusions about what true love is, perpetuating clichés and wrong ideas about love relationships.

English musician Sting romanticizes what sounds like an obsessive and controlling kind of love: a man telling a woman who's left him that she belongs to him and therefore, he'll be watching every move she makes. And if it was meant to express adoration (or a protest against surveillance technology, as has been said), it certainly sounds more like a stalker's song!

Misogyny and lustful content in songs are not exclusive to modern lyrics. We hear it in forms like reggaeton. Tango, which prides itself of its elegance and is widely socially accepted in Latin America, often carries a very macho tone.

It's difficult to judge the past with the level of consciousness we have in the present. The idea of a "loving" relationship in which men had the right to violate their women might have been acceptable in times past, but in most Western countries today we're clear it's unacceptable.

Examples of misogynistic or violent songs are Chris Brown's "Fine China" (2014), in which he refers to a woman as an irreplaceable, collectible object. Or the way in which in "It's So Easy" (1987), Guns N' Roses refer to a woman as a sexual object.

Mainly because they contribute to transmitting a misogynist and violent culture, I've become very critical of some "love" songs' lyrics, and not only modern ones. I deplore especially those that cross the barriers of respect for women, in which they're portrayed as objects of desire and targets for violence. As disgusting as these songs feel, unfortunately they sell. The sad thing is that they incite the expression of the brutal aspects that loom in human beings. These songs promote aggressive sex and sexual license that, in my view, are the product of misperceptions of love and the consequences of people's loneliness in a troubled world where true eroticism and true intimacy are fading.

Upon closer examination it's easy to see that these songs aren't meant to inspire or invite sensuality. Instead, they contribute to affirming male power. And if the content hasn't been censored it's mostly with the argument that they don't cause direct, tangible, physical harm.

Some would defend the popularity and acceptance of these expressions as the result of women's liberation, modern females becoming unafraid of rebelling against the establishment or of revealing themselves as rightful sensual beings. Of course, there are women who also criticize the ubiquitous eroticization or the expression of sexuality as mere penetration. We women demand to be appreciated in our wholeness instead of as a set of body parts (natural or artificially enhanced), void of both mind and spirit. But women who are against misogynistic and hypersexualized lyrics are often branded as prudish and repressed. In any event, in a consumerist and male-dominated society women's sexual expression tends to be conditioned, not spontaneous.

Misogynist content in rap music was studied by Ronald Weitzer and Charis E. Kubrin from the George Washington University. They found that of the 400 songs they examined, about 22 percent of the lyrics had some misogynist bias. Their content analysis identified five misogynistic themes: "(a) derogatory naming and shaming of women, (b) sexual objectification of women, (c) distrust of women, (d) legitimation of violence against women, and (e) celebration of prostitution and pimping."[139]

In February 2018, the *Harvard Crimson* reported that in the year 2017, for the first time in history, "hip-hop/R&B became the most consumed music genre in America with the average age of hip-hop listeners being the lowest of all major music genres in the United States." It's true, this music conveys powerful messages from the disempowered, but it also hits a young and vulnerable population with "misogynistic lyrics that hypersexualize and give little to no respect to women."

Rand Corporation researcher Steven Martino and his colleagues found a correlation between sexually degrading sexual lyrics and early onset of a sexually active life in adolescents, which goes along with an increased risk of unplanned pregnancy and sexually transmitted diseases.[140] Various reports show that early sexual activity has become problematic in the United States. However, studies show that teens often wish they had waited longer to start having sex and there is a correlation between unwanted pregnancies, sexually transmitted diseases, and the early onset of sexual activity. Martino's study concluded that "Reducing the amount of degrading sexual content in popular music or reducing young people's exposure to music with this type of content could help delay the onset of sexual behavior."

A misunderstanding of the meaning of sexual liberation might have led to a compulsive genitalization of erotic relationships, and

---

[139] R. Weitzer, and C. E. Kubrin. "Misogyny in Rap Music." *Men and Masculinities* Vol 12, no. 1: 3–29. (October 19). 2009)https://doi.org/10.1177/1097184x08327696.

[140] S. C. Martino, Marc N. Collins, M. N. Elliott, A. Strachman, D. E. Kanouse, and S. H. Berry, "Exposure to Degrading Versus Nondegrading Music Lyrics and Sexual Behavior Among Youth," *Pediatrics* Vol 118, Issue 2 (August 2006).

unsuccessful and transient attempts to connect with others. The culture in which we live is full of enticements that, even in the case of love, constantly direct us to "the next best thing."

Sex can turn into a currency with which to pay for minutes of ecstasy in company. Relationships are defined by lust, often restricted or delimited by the intensity of the physical pleasure experienced. But afterwards, there often comes a void and a search for more of that transient bliss, which can lead to promiscuity.

## Genitalization of love vs. tantric love

*So she thoroughly taught him that one cannot take pleasure without giving pleasure, and that every gesture, every caress, every touch, every glance, every last bit of the body has its secret, which brings happiness to the person who knows how to wake it. She taught him that after a celebration of love the lovers should not part without admiring each other, without being conquered or having conquered, so that neither is bleak or glutted or has the bad feeling of being used or misused.*

—Herman Hesse

Addictions to drugs and sex have much in common in the physiological (same neuronal circuits), emotional, and social aspects. Using drugs or abusing sex has little to do with the freedom to do whatever you please with your body, as some may claim. Quite the opposite, there's usually something lacking at the root of most addictive, self-destructive behaviors. And there seems to be an association between an imbalance of the dopaminergic centers and compulsive behaviors that seek a state of transient excitement and pleasure.

Of course, not everyone who has multiple partners suffers from a compulsion or a sexual addiction. Warning signs of an addiction include trouble limiting sexual activity at will, allowing sex to interfere with other activities (work, study, family life), dedicating a considerable number of hours to sex-related activities (pornography, masturbation, fantasies), not being able to control sexual activity in spite of being conscious of the consequences, and suffering from states of anxiety and irritability when there are no conditions to satisfy sexual urges.

Let's contrast some sexual issues, such as genitalization and addiction, with the ancient Hindu tantric tradition and practices. Tantra is an inclusive practice that means to expand and connect (it weaves many other practices and teachings, including yoga). Tantric sex continues to be a common spiritual practice in the East, but it's not usually advertised or promoted. According to this tradition, the divine couple, Shakti and Shiva (feminine and masculine principles, respectively), created the universe that gave origin to our world through their copulation and dance. The whole creation would then be the result of a loving and erotic act that is recreated through the union of a couple by following specific practices.

Like original yoga or Zen practices, tantra is not a religion. It's practiced for the purpose of obtaining enlightenment and its philosophy goes beyond the bedroom, expanding to all aspects of life. A couple mindfully focus on both full-body physical pleasure and on enhancing mutual connection through the synchronization of breathing, the exploration of erogenous zones, and the open expression of feelings and sensations. Tantric sex emphasizes sensuality, but by approaching the body as a sacred temple. During their interaction, the couple starts with the conscious intention of connecting with the vital energy (Kundalini) that's thought to reside in the first chakra. The belief is that the ascension of Kundalini through the main *nadis*, the energy channels known as Sushuma, Ida, and Pingala, running parallel to the spinal column, leads to a connection with the divine energy and an increase in the awareness of the meaning of human existence.

The tantric experience allows the couple to transcend desire. Tantric sex, whose practices are often trivialized in the West, is a practice that originated in India somewhere between 2,000 and 5,000 years ago. The purpose of tantra is to enhance well-being and spiritual transcendence. It doesn't aim at just achieving an orgasm by stimulating the genitals or erogenous zones, or even at reaching climax. Its goal is to give the union of two beings a holistic dimension: physical, emotional, mental, and spiritual.

## In search of equality

*Women are the only exploited group in history to have been idealized into powerlessness.*

—Erica Jong

I have seen dozens of couples in my practice who have unfortunately been victims of some form of emotional, verbal, or physical violence. The ensuing trauma leaves a long-lasting mark.

Any relationship must meet a set of minimum requirements to be healthy and to last. The most important of these requirements, I believe, are mutual respect, mutual support, and the certainty that you can be yourself, and express yourself in the whole dimension of your being, in the presence of the other. That is what I consider true intimacy.

If one partner feels they should thread lightly or otherwise something bad might happen (if they feel as if they're always walking on eggshells), this is a clear indication that the relationship has an abusive component and is, therefore, unhealthy. A situation in which one is not able to be oneself in the presence of the other and express oneself freely is inherently violent.

Duluth—a small community in Minnesota—found ways of holding batterers accountable and keeping victims safe. They believe the whole community needs to be involved in order to end domestic violence. The graph below (from their domestic abuse intervention programs) offers a list of conditions that a relationship must meet to promote equality and prevent different forms of abuse (emotional, economic, physical).

The prevalence of violence against women and harmful practices

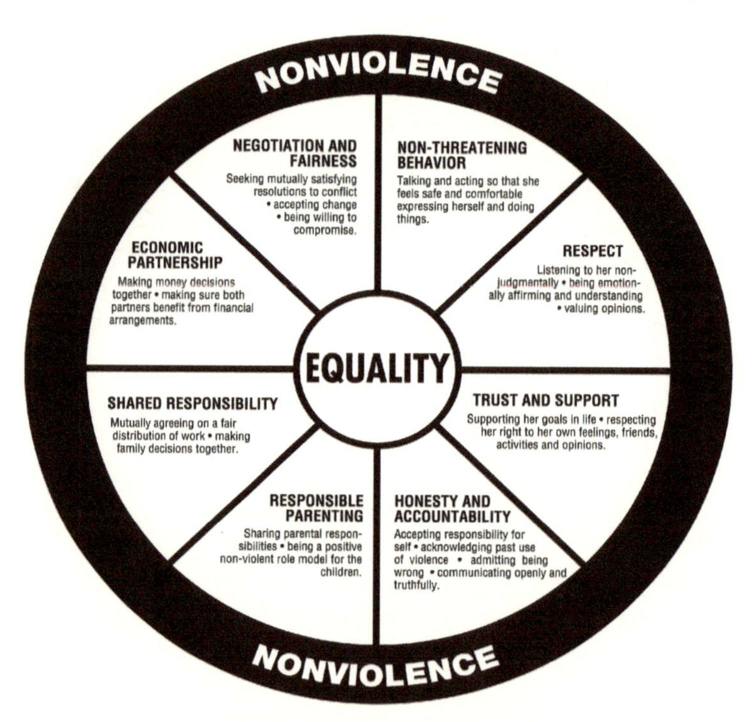


against them continues to be a serious concern, according to a 2018 United Nations statement: "It is estimated that 35 per cent [*sic*] of women worldwide have experienced either physical and/or sexual intimate partner violence or sexual violence by a non-partner (not including sexual harassment) at some point in their lives" (UNwomen.org). Unfortunately, many expressions of sexual violence persist and are even sanctioned by society in many cultures. However, there have been improvements to legal protections for victims of rape, especially in Western countries and in China. In 1993, the General Assembly of the United Nations addressed the need to eliminate violence against women, expanding its definition to include certain practices such as clitoris ablation, still commonly practiced among some adherents of the Muslim, Christian, and Jewish faiths. Even though female genital mutilation has been erroneously linked to religion, it predates Christianity and Islam. In the same session, the UN adopted the Declaration on the Elimination of Violence against Women.

In 2012, the UN declared February 6 as the day of zero tolerance for female mutilation. However, it's estimated that this practice still affects about 140 million women in the world. It's practiced among the Emberá tribe in Colombia, where it's illegal, as it is in most European countries. In August 2016, the Pan-African Parliament banned female genital mutilation in its fifty member states and it's currently taking measures to eradicate the procedure. Worth mentioning that most Muslim theologians consider ablation an unnecessary practice and contrary to true Islamism (which recommends female circumcision instead).

### #Nomore!

The hashtag #Nomore became a unifying symbol expressing support for ending domestic violence and sexual assault.

In 2017, newspaper reports by Jodi Kantor and Megan Twohey in the *New York Times* (*NYT*) set off a hailstorm of lawsuits against famed film producer Harvey Weinstein for sexual harassment. His behavior had been an open secret in Hollywood for three decades, and the first woman who dared to report him to the police in 2015 was ridiculed and defamed. After the *NYT*'s publication, actress Alyssa Milano, best known for her roles in the TV shows *Who's the Boss* and *Charmed*, posted a message on Twitter calling on all victims of harassment or any form of sexual violence to identify themselves in the social media with a #MeToo. It was the beginning of a shower of allegations against powerful personalities and celebrities (politicians, television presenters, producers, actors) and the origin of an increasingly strong movement in the United States against sexual harassment and sexual assault.

In just a few hours, thousands of people had made public the experiences of harassment or sexual violence of which they had been victims. Many celebrities became spokespeople for this movement and have used their highly visible appearances in events such as the Oscars and Grammys to make statements advancing the rights of women, denouncing sexual abuse, and calling for the creation of a safer world.

After receiving the Cecil B. deMille Award in 2018, Ophra Winfrey delivered an acceptance speech in which she denounced "a culture broken by brutally powerful men." She pointed out that telling the truth out loud is the most effective tool there is to stop abuse. "For too long," she said, "women have not been heard or believed if they dare speak the truth to the power of those men. But their time is up."

The United States seems to have finally broken a silence often coerced by perpetrators, sometimes self-imposed by the shame victims experience after being harassed or assaulted. Many find it difficult to understand why victims (not just women) keep silent, sometimes for decades before daring to denounce their abusers. It's not only because of the shame they endured but also because the conditions that would allow such victims to speak up didn't exist at the time they were abused. Women have been persuaded that "a guy is a guy," and he supposedly can't contain his sexuality. Women have been taught to be quiet and to endure. For a long time, there was a stigma attached to the loss of virginity or the free expression of female sexuality (it still is in many countries).

Women's silence unintentionally kept the abusers safe and free of legal consequences. The deplorable truth that now comes to light is that for years, women and other minority groups (LGBTQ, blacks, Indians) have suffered not only sexual abuse but also domination by individuals belonging to majority groups.

Though the meaning of love has evolved on par with women occupying their rightful place in society, and though they're speaking up and resisting the harassment and sexual abuse suffered for centuries, the word *love* is still ominously trivialized. And yet, in other ways, the word has also been hyperbolized. Breaking stereotypes is not easy.

## Loving from our strength

*On the day when it will be possible for woman to love not in her weakness but in her strength, not to escape herself but to find herself, not to abase herself but to assert herself—on that day love will become for her, as for man, a source of life and not of mortal danger.*

—Simone de Beauvoir

Women's and men's positions in society have made them vulnerable when it comes to relationships.

True love is not about sacrifice. Simone de Beauvoir's quote above is brilliant: a healthy relationship should lead to affirmation, growth, and discovery of ourselves.

Our misconceptions about love, for example, prevent us from spotting the red flags indicating we're entering into an abusive relationship. This blind spot in our vision contributes to the cycle of abuse: after an episode of abuse, the abuser apologizes, makes promises, and brings the relationship back to an idyllic state of roses and chocolates. So, the victim surrenders again to the charms of the partner (charms that might be real). Then, little by little, the relationship gradually slips back again into abusive interactions. A victim often believes the abuser. At first, she just doubts herself, makes excuses: it was probably a misunderstanding, or he was having a bad day. In time she starts asking what's wrong with her. When the abuser strikes, she honestly wonders if it was her fault. She's seen him at his best. She reasons that her partner is probably only out of character for a moment and that she can win back his affection.

I frequently counsel women who've been victims of some form of abuse—verbal, emotional, physical, financial, or sexual. They've been stripped of personal power. The reasons why they enter into abusive relationships are diverse, but sometimes it's for what one of my patients called the fairy tale. She referred to an idealization of romantic love that seems to come from childhood stories and movies with happily ever–after endings. She alluded to the notion that we must be in a relationship to be happy. She also referred to a common misconception that a "good provider" is the direct equivalent of an "ideal partner." Besides, she talked about another widespread belief—that since all relationships go through hurdles, there's no real need to address conflict or mistreatment; grief will ease with time.

The above assumptions place the woman in a passive, weak, and dangerous position.

Verbal and emotional forms of abuse are more persistent and frequent than physical or sexual abuse, but they go underreported. Among other reasons, this is because the perpetrator manages to cut the victim's access to her support systems, debases her with constant criticism, and restricts her access to financial resources. Abusers seek to maintain control and power. Abused people's self-esteem becomes so compromised that in most cases they no longer feel capable of breaking their bond and dependency. This problem is serious, persistent, transcultural, and affects both heterosexual and homosexual couples.

Based on "Findings from the National Intimate Partner and Sexual Violence Survey," the nomore.org site reports that:

- More than one in four women and one in nine men experience violence from their partners in their lifetimes.
- More than one in three teens experience sexual or physical abuse or threats from a boyfriend or girlfriend in one year.
- More than one in five women are survivors of rape.
- More than one in three women and one in six men have experienced sexual violence in their lives.
- More than one in four women and one in six men were sexually abused before the age of eighteen.

In the United States, up to 80 percent of victims of physical abuse end up with symptoms consistent with the diagnosis of post-traumatic stress disorder.

An online survey launched in January, 2018, by Stop Street Harassment found that 81 percent of women and 43 percent of men experienced some form of sexual harassment during their lifetimes.

Around 25 percent of homicides in the United States involve spouses, sexual partners, or love rivals. Men have reported almost four hundred thousand cases of harassment each year, while almost two million women are physically attacked by their husbands annually. Every year, close to one million women in the United States report harassment by their ex-partners.[141] The United Nations Office on Drugs and Crime reports that in all countries where data is recorded, statistics show most cases of violence are motivated by jealousy, and the abuser is more often male (87 percent of cases).

In many of the cases where violence is mutual, women have initiated it. Although the aggression perpetrated against a woman is automatically interpreted as gender-based violence and is considered the consequence of an unequal level of power, men are also often the victims.

Power assumes the right to control others. Strength assumes no such power, and instead takes full responsibility for the choices made. Abuse Causes emotional wounds that could leave marks on the victim for life if help is not promptly sought. Many women feel trapped in a relationship; others do separate from their partners but have a hard time recovering emotionally. Very often victims enter into new abusive relationships, and still others become aggressive themselves and even abusive. Some women who have grown in strength despite an adverse environment find the courage to leave their abusive partners and rebuild their lives. A few of

---

[141] Numbers from the Bureau of Justice Statistics and the National Coalition Against Domestic Violence.

them become strong enough to reenter the same relationship, but on their own terms.

Let me also bring your attention to the fact that more than 70 percent of love suicides in the United States are committed by men. Relationship failures can be devastating, and they often lead to depression. Rejection is so difficult to take that it's a common cause of suicide.

*A patient who, after forty-plus years, separated from a husband she described as abusive (verbally and emotionally), admitted the husband back into her life after two years of separation. She did so very slowly and fearfully, but she reestablished boundaries, reaffirming her right to have her own space and indicating which of his behaviors were inadmissible.*

*"When alone, I don't have to listen to your constant criticism," she firmly told him one day when she was riding in their car. He was staying at their summer home, as agreed when they'd separated. She'd married very young and had soon developed a dependency on her husband. She hadn't foreseen the difficulties she would encounter once she started to live on her own, but she found she could fare well by herself, and after a year she felt like a new, empowered person.*

*"For a while I wasn't moving in any direction," she told me, "but therapy helped me be who I wanted to be." She learned that the key was to be able to shape her own destiny. "Most women live in a fairy tale and do not work on themselves," she said. "They get caught up in their own dance."*

## Who is your friend?

If I like myself, accept myself, and love myself without reservations, I should be able to love others totally and unconditionally. Most of our grudges, resentments, and regrets are rooted in experiences with people we loved but who humiliated us. We keep them at a distance; we fight with them, trying to prove we're right. But in most cases, that fight ends up excluding love and cracking or ending relationships. Could we forgive them? Both body and mind go on alert when we've been hurt. Our fight-or-flight response becomes easily activated.

My Qi Gong teacher explained how Chinese believe that there are three main types of people in our lives with whom we're linked by energetic cords. First are those who love and accept us exactly as we are, to whom we can bare our souls without reservations. We could call this group our real family. We must cultivate and protect these relationships.

In a second group are those people whose company we enjoy but whom we don't know deeply enough. We consider them friends or acquaintances. We're still testing the waters with them.

And finally, there are those people who often hurt us, criticize or cheat on us, abandon us in crucial times. We might have family ties with them, but they often reprimand us, threaten and punish us, or compete with us. Even worse, they're people who have humiliated us. Consider these group of people toxic. Chinese say we must cut the energy cords that bind us to these toxic people so that we can pull off the power they exert over our lives. Avoiding relationships with toxic people does not exclude loving or forgiving them, but we must not forget how they make us feel. And, we shouldn't offer them any opportunities to abuse us again.

While forgiveness is critical to our physical and emotional health, so is protecting ourselves physically, emotionally, and energetically from those who have the potential to hurt us.

Ideally, our behavior would be guided by this question: What is the most loving thing I can do in this case, for myself and for the other? We could then consider the possibility that when we feel angry with others, we're actually angry at our own inability to understand and love others unconditionally.

# PART SEVEN: We are One

*Quantum theory thus reveals a basic oneness of the universe. It shows that we cannot decompose the world into independently existing smallest units. As we penetrate into matter, nature does not show us any isolated "building blocks," but rather appears as a complicated web of relations between the various parts of the whole. These relations always include the observer in an essential way. The human observer constitutes the final link in the chain of observational processes, and the properties of any atomic object can be understood only in terms of the object's interaction with the observer.*

—Fritjof Capra

## Transcendental love

The individual's drive toward individuation, which began in early childhood, continues during the teen years with the search for an identity, and it extends throughout our lives. It's as if we had hung the question "Who am I?" over our heads. We're steadily moving forward, trying to reaffirm our identity, which is made up of ethnic, geographical, family, gender, and ideological components. It's also a long quest for meaning and purpose. And, in the last few lustrums of our existence, when we thought we were really close to finding all the answers, our life suddenly modifies the questions. In childhood, adolescence, and the beginning of our adult lives, we kept our gaze straight ahead, on the horizon. Then we looked upward, determined to reach the top. But there comes a moment in life when we find ourselves at the pinnacle of life and looking down. Life brings us to a point where our impetus calms down, but with a constant awareness of our now inevitable decline and the view of the final stretch ahead. Some of us try to resist or deny the foreseeable fate. Others simply acknowledge and accept their mortality and prepare for the conclusion of life.

Following the generativity stage described by the psychologist Erik Erikson, there comes the eighth and last stage of psychosocial development, which he characterized as "integrity versus despair."

According to Erikson, this stage begins around the age of sixty-five, and it's a time for integration and a desire to take stock of one's life. This stage is characterized by the exploration of the common thread that's given meaning to our lives. The main risk of this stage, despair, comes from feeling that our life has not been worth living and hasn't been productive enough, or that we haven't fulfilled our purpose. Despair also comes from fear of loneliness, physical decline, and death.

Our opportunity, on the other hand, would be discovering or further cultivating a form of transcendental love and getting ready for the great departure. By this age, egocentrism and vanity must already be minimal; physical appearance usually is not as much of a concern. Even if the mind is still intact, the body has already betrayed us, giving us wrinkles and aches if not maladies. Maybe our main existential questions now revolve around deciphering our essence. It's likely that people at this stage look back, trying to understand the role they've played in the lives of others and what contributions they've made to the world. Then again, they might choose to disregard all explanations as unnecessary and merely accept what is.

In any event, when reaching the so-called golden years, ideally, we would have developed our full capacity for empathy.

In old age we're more than packing for the big departure. If we're able to retire and health and finances allow, we'd be able to enjoy some free time to scratch things off our bucket list or to be more actively dedicated to social service. Egocentricity should be minimal by now, and we'd be more willing to listen to others. With the mind free of other mundane matters, a spiritual dimension is often discovered (or rediscovered), and we should be able to more easily detach from the bonds of this world.

This is a period in which, as in the poem "Desiderata," we must "take kindly the counsel of the years, gracefully surrendering the things of youth."

As in Greek mythology, the term golden age is used to describe a last stage, in this case of our lives. It also refers to a time in which we have achieved a certain harmony, stability, and prosperity, and it's no longer necessary to work to make a living. Ideally, at this age we retire and live from a pension. Our lives are no longer upset by substantial changes, and we've reached the best conditions to live in peace.

Regrettably, some people reach their final stage full of resentment and bitterness, have trouble accepting life as it is, and feel incapable of forgiveness. Also, it inspires compassion to know that, according to the 2005 United Nations Habitat global survey, around a hundred million people were homeless worldwide, and as many as 1.6 billion people (20 percent of the world's population) lacked adequate housing.

In 2014, the US Department of Housing and Urban Development reported that about 306,000 people over fifty were living on the streets in the United States, representing 31 percent of the nation's homeless population. More recent statistics do not report on homelessness above the age of fifty.

Several studies show that the quality of life (QOL) of a person in their last years strongly depends on previous lifestyle, income, education, and family support. Spirituality and integration into the community could be as significant as access to health, according to several studies carried out in countries such as India and Bhutan. However, many scientists consider that a factor as tricky to define as spirituality should not be taken into account when measuring quality of life.

The salience of helping others (prosocial behavior) is seen more frequently among older adults than among young people,[142] and these behaviors in turn have a correlation with a state of greater emotional empathy and greater cognitive empathy.[143]

A (qualitative) study with healthy nursing home residents in Norway[144] showed that individuals who could find meaning, despite hardships or illness, withstood life challenges better than those with a low sense of meaning. They listed four key meaning-making experiences: 1) physical and mental well-being, 2) belonging and recognition, 3) personally treasured activities, and 4) spiritual closeness and connectedness.

In the United States, at least half of the people of retirement age adhere to a religion,[145] many of them attending non-denominational churches. A higher percentage of seniors attend church services than in any other age group, and churches have become one of the most

---

[142] J. A. Sze, A. Gyurak, M. S. Goodkind, and R. Wl Levenson, "Greater Emotional Empathy and Prosocial Behavior in Late Life," *Emotion* Vol 12(5) (October 2012): 1129–114.

[143] J. N. Beadle, A. H. Sheehan, B. Dahlben, and A. H. Gutchess, "Aging, Empathy, and Prosociality," *The Journals of Gerontology*, Series B, Volume 70, Issue 2 (March 2015): 213–222.

[144] J. Drageset, G. Haugan, and O. Tranvåg, Crucial Aspects Promoting Meaning and Purpose in Life: Perceptions of Nursing Home Residents," *BMC Geriatrics*, BMC series (2017): 17–254.

[145] M. TsinChiu, B. Ofstedal, F. Rojo, and Y. Saito, "Spirituality, Religiosity, Aging and Health in Global Perspective: A Review," *SSM—Population Health* Vol. 2 (December 2016): 373–381.

important sources of support besides family. This could be interpreted as indicating that approaching the end of existence moves people to seek a more spiritual life.

We held Reiki circles regularly at our holistic center in Bonita Springs, Florida, and they were open to the community. These free gatherings, held twice a month, were attended mostly by people over fifty who, after participating in a short meditation session, received a brief relaxing treatment. They kept attending these circles, so that they could feel again the serenity that many of us experienced while being together.

As I write these lines, I realize how difficult it is to put into words this concept of transcendental love, since it refers to experiences that go beyond everyday life and combine mental states of loving presence and connectedness. Readers who've meditated or practiced mindfulness might more easily understand what I'm trying to describe. My personal feeling is that transcendental love is about an abstract experience of being fully present, where no worries exist, and the consciousness of separation disappears. When we achieve such a state, suddenly we're no longer strangers to each other.

Before becoming president of Egypt, while he was a military officer during the Second World War, Anwar el-Sadat conspired to free Egypt from English domination. This cost him his freedom; he was imprisoned and placed in solitary confinement. In his autobiography, he revealed that during the time he was secluded, he was able to transcend the confines of time and space. In that state, he felt connected to the universe, and his tolerance for his circumstances increased. He proposed putting aside feelings of hatred and revenge, convinced that the just would triumph. "I discovered the beauty of love," he said. Even his concept of God changed, and he envisioned no differences between the Gods of Christianity, Judaism, or Islam.

Psychologists are increasingly interested in studying this form of transcendent love, in which we experience mental states characterized by feelings of interconnectedness with others and with everything that exists. Fromm examined a concept of God and how it has evolved along with society and how it makes our cultural differences evident: in the West, love of God is more than anything a mental experience, Fromm says, while in Eastern religions it is an intense affective experience of unity. Contemporary authors such as Caroline Myss, Eckart Tolle, Neal Donald Walsh, and Elizabeth Gilbert have described how, in the midst of intense psychological despair, they've experienced what the Vedas called "cosmic consciousness," moments in which the mind looks at itself. Such moments, they concur, contribute to transforming our perception of the world and the course of our lives. In *The Tao of Physics*: *An Exploration of the Parallel between Modern Physics and Eastern Mysticism* (Shambala Publications, 1976), Fritjof Capra describes one such experience. He entered a meditative state in which he perceived what he calls a "cosmic dance of subatomic particles."

These states, which could be considered mystical, are not uncommon among those who practice mindfulness and other forms of meditation that include positive feelings (generosity, compassion, gratitude, amazement, tranquility).

Transcendental meditation (TM), a practice that includes repeating a mantra, can be combined with mindfulness, both while the person is awake and during sleep, seeking an integrated state, known as cosmic consciousness (Vedic tradition). This inner state coincides with the appearance of alpha brain waves in an electroencephalogram (typical of a person who is relaxed but still awake) and delta waves (typical of deep sleep). It is said to contribute to increased brain integration and emotional stability, and lowered anxiety.

During the so-called golden years, many people seem more inclined to service, which is a manifestation of the experience of unconditional and sometimes transcendental love. Statistics show that, in the United States, women volunteer at a higher rate than men, and the greatest number of people who do volunteer work are over forty years old. If seniors volunteer in fewer numbers perhaps it's mostly due to the health limitations that often come with age.

We already mentioned that Piaget, who coined the term "logical egocentrism" (seeing the world from your perspective, considering yourself the center of what happens), believed the egocentric perspective begins to develop at age two and evolves over the course of our lives. Adolescents, who are developing logical and abstract thinking, can put aside egocentricity and look at the world with critical and transforming eyes. They often have moments of extreme lucidity, in which they can see the world more objectively, analyze social phenomena and its causes, and adopt their own views about how to face them. Almost all of a sudden, they see parents critically, confront those responsible for the ills of society, and feel motivated to be and do better than their predecessors, pledging to correct the course of the world. This aspiration propels them forward and gives rise to their altruism, a desire to be of service, and the experience of disinterested love toward others.

In "Millennials Are Effecting Change with Social Responsibility" (*Forbes*, August 2017), Wes Gay writes, "This generation is passionate about social causes that benefit the greater good, whether it's a nonprofit charity or an altruistic company like TOMS (toms.com).... As a whole, millennials tend to be generous with their time, money and influence.

They freely use their social media platforms to raise awareness and money for causes important to them."

Our paths don't seem very different from that of Siddhartha, the fictional character on the homonymous novel by the German novelist Herman Hesse (Cambridge, MA: New Directions, 1951). In his quest for enlightenment, Siddharta goes through different stages that have been characterized as the stage of the mind, the stage of the flesh, and the stage of transcendence.

Raised as an overprotected child, prince Siddhartha rebels against his father, who has provided for his every need but has also blocked his access to the reality of the world. He leaves the Brahmin's palace to seek his own way of achieving lasting satisfaction. Siddhartha tries different paths. He becomes a beggar first and among other practices, he learns from the ascetic Samanas to meditate, be patient, and fast. But he leaves their company when he concludes that their methods led him into a form of escapism and self-denial. He then decides to listen to the Buddha only to reaffirm his belief that no one can teach us our way to enlightenment because we find it on our own. For a time, he becomes a pilgrim. He learns that if he has clear goals and faith, fears do not get in his way. At a given moment, still looking for the meaning of life, he leaves that mystical path (the path of the mind) and meets the courtesan Kamala who demands him to learn the ways of the city before she teaches him the art of physical love. With her help and connections, and since he had reading and writing skills, he becomes a successful merchant and obtains material riches. Then one day, he painfully wakes up. Before Kamala dies from a snakebite, she reveals to Siddhartha that her 11-year old is also his son. When Siddhartha sees his son grieving for his mother, he tries to protect him, but a wealthy life has spoiled him, and his father's kindness overwhelms him. After a while, his son runs away and, after an unfruitful chase, Siddhartha realizes he can't prevent his son from following his own path. The wealth he has obtained seems not enough to make him happy. Was it worth leaving his father's house and abandoning both his best friend Govinda and his quest for enlightenment? He enters a state of despair and even considers ending his life but, encouraged by a ferryman, opts for a contemplative life instead. And, unexpectedly, he has a transcendent experience. While meditating by the river, he perceives a sound he remembers well: OM, the primordial sound, with which he reconnects with Brahman (in Brahmanism, the reality par excellence, the unity of it all). Siddhartha achieves enlightenment when he had already given up hope.

There's no unified way to rise above our ordinary limitations to find transcendence, but we can certainly seek it. A transcendent love for life, for others, for everything that exists, is par excellence an impersonal and unconditional type of love.

## Voices of love—Love as service

*Justice will not come to Athens until those who are not injured are as indignant as those who are injured.*

—Thucydides

Thanks to advances in telecommunications over the last hundred years and widespread access to the internet, the sound of the voices of love throughout the world are now within our reach. Voices that speak of sacrifice, of service, of compassion. But the volume of these voices is often muffled by the noisy drums and cymbals coming from the show business world and from deceptive propaganda. Marketing drives us to seek happiness not in our connection with others, not in service to our fellow human, not in the expression of our talents or creativity, not in the search for purpose and meaning, but in the possession of more and more stuff: Be happy; buy an iPad! Many people go shopping in an unconscious effort to mitigate their sadness or loneliness. People buy a lot more than they need and are still not satisfied.

Between 2011 and 2014, approximately one in six Americans was taking antidepressants (a quarter of them had taken the meds for ten years or more), while three decades ago, only one in fifty took them. One reason may be that depression is diagnosed more often. Another reason is that the quality and availability of drugs has increased. Also, doctors are now more open to prescribing pharmaceuticals, and patients are more willing to take drugs than commit to psychotherapy. However, studies show that therapy provides better results in the long term.[146] However, what's striking is that the percentage of people feeling depressed is so high. Studies show a connection between a poor diet, nutritional deficiencies, and depression, but that's only part of the picture. Not taking care of the body, not exercising, and not eating healthy are symptoms of lack of self-love.

[146] J. Siddique, J. Y. Chung, C. H. Brown, and J. Miranda, "Comparative Effectiveness of Medication Versus Cognitive-Behavioral Therapy in a Randomized Controlled Trial of Low-Income Young Minority Women with Depression," *Journal of Consulting and Clinical Psychology* Vol. 80 (6) (Dec. 2012): 995–1006.

Philosophers have postulated that depression and feeling an existential vacuum in life has become a contemporary evil. For the older generations it's always tough to live in a world that no longer holds the traditions and values that made sense to them. For the youngest, it's tough that there are fewer inspiring leaders to look up to.

Not only have our days turned into a race that give us no time to appreciate life, but in their search for happiness, people have focused on acquiring professional success, fame, and money. There's not much time available to meaningfully connect with loved ones or for being of service to others. *Nonforprofitquarterly.org* reports recent declines in national volunteer rate. The report says that the 2015 rate is "the lowest rate measured since the CPS began conducting annual volunteer surveys in 2002."

Increased access to news about mass killings, terrorism acts, exodus, or famine overwhelm us, and we might opt to ignore them, until these events become alien to us.

Many twentieth-century leaders inspired us to transform the world. Financial success wasn't always the first priority. In the United States of the first half of the twentieth century, volunteering was an essential value. Many people dedicated their lives to service, to promoting solidarity and justice. And though we notice the noisy fanfare of today's famous billionaires, the old voices are still rumbling, new voices are starting to become louder, and more youth want to transform the world.

Let's take some time to honor the memory of some those voices from the past:

Florence Kelley, a social reformer of the early twentieth century fought against inhumane working conditions in factories. She opposed child labor and managed to introduce legislation for a minimum wage in the United States.

The ideas of the pedagogue John Dewey significantly influenced the process of the democratization of schools through progressive education that postulates that a hands-on approach facilitates children's learning. According to Dewey, teaching was more effective when offering meaningful activities and allowing democratic participation in the classroom.

Dr. Martin Luther King Jr., who in the 1960s led one of the most important movements for civil rights in the United States, advocated nonviolence and promoted equality for all. Through protests and marches, his leadership led to the passing of laws against racial discrimination. His dream of equality is still shared by many.

Nelson Mandela, a political prisoner for twenty-seven years, left prison in 1990 to negotiate the end of the apartheid, and he was eventually elected the first black president of South Africa. For many, he is still an example of compassion and justice.

Mandela said, "Our human compassion binds us to one another—not in pity or patronizingly, but as human beings who have learnt how to turn our common suffering into hope for the future."

It seems that as we've become more materially comfortable, although we're moved by positive news or inspiring stories of sacrifice or by the condemnation of situations that require urgent remedy, we have become slower to take action, and we too often fail to make the effort to put our empathic capacity into practice.

On September 2019, thousands of youngsters around the globe took the streets to protest climate change. Thousands of young people have rallied in the streets of the United States seeking a ban on access to assault weapons. They're giving us a light of hope with their determination. The youth marching in Colombia for the right to education and the women raising their voices against gender violence inspire us.

Last century saw leaders like Mao Zedong, who had dreamed of becoming an educator and wanted to be remembered as a teacher but ended being the inspiring leader of the People's Republic of China (1949–1976). Mao told his people, "We should be modest and prudent, guard against arrogance and rashness, and serve the Chinese people with heart and soul…."

Talking about Mao: at the intersection of Guy and de Maisonneuve streets in Montreal stands the statue of a man whose life had already inspired me before I came to study at Concordia University in 1989. It's the image of doctor Norman Bethune, who left Canada in 1937 for the last time to put his professional talents at the service of the People's Army of China, which, led by Mao, was fighting against a Japanese invasion.

Bethune was a thoracic surgeon at the Sacré-Coeur Hospital, in Montreal, where he had made novel contributions to his profession (he invented the Bethune Rib Shears, still in use), when the news of the Spanish Civil War motivated him to enroll with the International Brigades to support the forces fighting fascism.

Thanks to Bethune's initiative, a mobile service was created in Spain, the first of its kind, to transport and deliver blood for transfusions to the battlefront. France provided the vans that would make this service possible, a service that preceded the creation of mobile army surgical hospitals (MASH). His unit was the first to transport blood—donated by civilians—to front-line hospitals. He also organized a rapid transport of type O negative blood, used when the recipient's blood type is unknown.

After the historic massacre of Malaga (February 1937), which caused the greatest exodus during the confrontation, Bethune ran to Almeria to care for the refugees.

He left a written report of the horror he had to witness: "A row of 30 kilometers of human beings, like a gigantic worm with innumerable feet that raises a cloud of dust, extended until beyond the horizon…they hung hungry in the fields, gripped, moving only to nibble some grass. Thirsty, resting on the rocks or wandering trembling without direction.… The dead were scattered among the sick with eyes open to the sun."

When Bethune witnessed this horror, he emptied the ambulance that transported instruments, equipment, and blood, and used it to mobilize the wounded. He reported that they transported up to forty people at a time. Today in Malaga, Bethune is recognized as a hero, and statues and memorials of him are found throughout Spain.

After serving in the Red Army for almost two years, Bethune underwent surgery, and died of septicemia in *Tanghsien, Hopei,* China, in 1939. Mao acknowledged his contributions: "Now we are all commemorating him, which shows how profoundly his spirit inspires everyone. We must all learn the spirit of absolute selflessness from him. With this spirit everyone can be very useful to the people. A man's ability may be great or small, but if he has this spirit, he is already noble-minded and pure, a man of moral integrity and above vulgar interests, a man who is of value to the people." Bethune is one of the few Westerners to whom a statue has been erected in China.

Also, worth remembering is Dr. Albert Schweitzer, who believed the purpose of life was to serve. He defined service as practicing compassion and having the will to support others.

Schweitzer was a Franco-German theologian, philosopher, musician, and doctor who won the Nobel Peace Prize in 1952 for his philosophy of reverence for life. He said, "Good consists in maintaining, assisting and enhancing *life*, and to destroy, to harm or to hinder *life* is evil."

His philosophy was expressed tangibly in 1911 when he and his wife founded the hospital that today bears his name, in Lambarene (Gabon, West Central Africa). There he attended hundreds of patients who came from everywhere. His purpose in life was to alleviate human suffering, and to this end, he studied medicine. He said his missionary work was a response to the call of Jesus Christ, but also as a small colonizer's retribution for the injustices and cruelties white colonizers had committed in Africa.

The Dalai Lama and Pope Francis in this century are voices of love as well. The Dalai Lama says, "World peace must develop from inner peace. Peace is not just mere absence of violence. Peace is, I think, the manifestation of human compassion."

And Pope Francis recommends having compassion for those who suffer, not just pitying them. He said, "Compassion is a feeling that involves you, it is a feeling of the heart, of the entrails, that involves you completely. Compassion is not the same as grief, it is not the same as saying: what a shame, poor people. No, it's not the same. On the contrary, compassion makes you get involved. It is a 'share with.' That is compassion."

Compassion demands, Francis explains, approaching the one who suffers in order to feel their reality.

None of the leaders mentioned above was perfect. History will judge their achievements and mistakes. However, they must be honored and listened to for the great endeavors they undertook. Their sacrifices contributed to transforming the lives of millions of people. These are the big questions for our old age: What has been my contribution? What has been our most loving act? How many lives have we touched?

## Unconditional love

*We live in the world when we love. Only a life lived for others is worth living.*

—Albert Einstein

Human offspring are among the most vulnerable creatures in the animal kingdom and require attention and care for a longer time than any other primate, or any other mammal, for that matter. Psychology professor Dacher Keltner points out that it's precisely this fact that made us a species especially inclined to extend the care of our kids for a lengthy period of time. This also explains the creation of social networks that facilitate the kind of group cooperation needed to take care of the children.

Specific areas of the brain light up when we experience unconditional love: the middle insula, the cingulate cortex, the superior parietal lobe, the caudate nucleus, and the ventral tegmental area (reward system). These same areas usually activate when experiencing maternal and romantic love.

Unconditional love can be either personal or impersonal. We can develop a strong affection for our children, students, nieces, nephews, or pupils. It's an unconditional type of love but of a personal nature. When we love unconditionally, we offer unlimited support, we fully accept the other, empathy flows naturally, we're inclined to forgiveness.

An ideal teacher, like a parent, usually feels empathy and compassion for their students, loves them for what they are, and other than academic performance or a certain behavior, the teacher expects nothing in return. However, this is really an asymmetric kind of love. The child loves his parents, guardians, or teachers mostly because he needs them. With luck and time, the child's love can be transformed into a combination of unconditional love and a sense of duty. Disciples could also develop a devotion for their teacher and an altruistic inclination to spread their teachings.

However, a parent or guardian's love might be limited because of its personal nature. There is an implicit condition needed for this love to happen: to be good parents or good teachers.

Impersonal unconditional love is what allows us to give of ourselves to others. It may transform us into true artists offering our art to the world. It might make us real teachers that generously share our knowledge with all. Unconditional love is not utilitarian: it's love offered regardless of circumstances.

When children begin to awaken their love for siblings and friends, they learn the value of loyalty, an essential component of unconditionality. Many childhood friendships are so strong that they last a lifetime. Fraternal love allows us to build bridges with other people in the community and to be of service to others. It's also an ingredient of the bond we develop with work peers and grows when we develop projects with a team. The characteristics of fraternal love are companionship, being able to count on the other (unconditionality), and trust.

Altruism, giving freely, would be the ultimate expression of unconditional love. Developing and practicing compassion will make us more altruistic.

As the director of the Mental Health Research Unit and a professor of clinical psychology at the University of Derby in Great Britain, Paul Gilbert,[147] founded compassion-focused therapy (CFT) to promote mental healing through encouraging patients to be compassionate to self and others. Gilbert considers relationships powerful regulators of our physiology and that taking care of others brings about major changes in the physiological regulation of our emotions. CFT comprises basic principles of neuroscience and Buddhism. Gilbert has published and

---

[147] Watch video: Paul Gilbert (2012), *Empathy and Compassion in Society* (YouTube).

edited twenty-plus books on compassion and related topics and has trained dozens of therapists. CFT is taught around the world.

Automatic responses to stress can take over our lives. However, we have the option to learn and mindfully choose how we respond to challenges that threaten to take us out of balance. A positive outlook, receiving support, but above all, the satisfaction we derive from connecting and being of service to others activates the production of feel-good neuromodulators and hormones.

Once the heart opens, we're no longer limited by a self-preservation instinct. With luck and persistence, we learn to listen to our bodies, we honor intuition, and we conquer a transcendence that allows us to feel that we're truly made of stardust, that is, that we all have the same origin; we're all made of the same matter and the same energy. We're all connected, and our apparent diversity merely expresses the creativity of the universe. Feeling connected makes it natural for us to show unconditional love and the joy experienced in the act of giving is not dependent on the behavior or reciprocity of others.

When providing a service, we must understand that we're not responsible for another person's happiness. Many disappointments in relationships and life come from the erroneous notion that others could make us happy or that we can get people to value us for our unsolicited services and generosity. In this case, we'd be coming from dangerous hopes: it's easy to feel disappointed if we give while expecting reciprocity or gratitude. The love you give doesn't always bear those fruits.

Is love between couples unconditional? Ideally, unconditionality becomes an ingredient of a solid relationship. However, it's not uncommon that when a person gets into a conflict with their partner or considers splitting up, they'll bring up all they have done for the other, expecting to be appreciated or valued. Not a sign of unconditionality. But it's all part of the learning curve.

A necessary condition for the expression of unconditional love is freedom. Not freedom as selfish individuality but as self-consciousness. As allowing the other to fully express who they want to be. But, according to Zygmunt Bauman, our love has been reduced to little. The last generations have grown perhaps with much more sense of individual independence than of duty, and often with a low tolerance to frustration that makes it difficult to defer gratification. In a dilemma between having freedom and security, contemporary humans have introduced changes in their lifestyles that have made love impermanent and conditional. And, if relationships revolve around our virtual communication, how much room for unconditional love is there?

## One for all...

*Love does not consist in gazing at each other, but in looking outward together in the same direction.*

—Antoine de Saint-Exupery

There is a feature that has become more and more evident to me in my last trips to Colombia. In more individualistic societies, sometimes it feels people don't give a damn. Although I have seen people cooperating in cases of crisis, in everyday life, everyone seems busy with their own affairs, and if you offer help to someone, let's say someone who's overloaded with packages or has dropped something on the floor, the most likely response is mistrust or, "I don't need your help, thanks."

*When I travel to Bogotá, which is not very often, my favorite meeting place with relatives and friends is a shopping center built on a beautiful historical hacienda called Santa Bárbara-Usaquén. The center's main attraction for me, apart from its colonial architectonic features, is a plaza, just a couple of steps outside the mall, which during the weekend hosts a market for multicolored handicrafts. Vendors from all over the country display art and cachivaches (creative artifacts) there. I prefer artisanal objects and domestic items over the plastic imported gadgets that have invaded Colombia's commerce.*

*On my last trip, I looked for a watercolor painting. After discussing the price of several pieces with one painter, I told him I'd come back later. I wanted to look at other artists' works and compare quality and prices. I returned later, wanting to buy something from him, but the painter was no longer in his booth; it was unattended. I asked a neighbor vendor if she knew where the man was. She made a sign to another vendor who came to see how she could help. One of them happened to have the phone number of the artist, who was gone for lunch. They wanted to take care of the business themselves, to make sure the sale happened, they said. But I needed the artist to provide me with packaging and a certificate of authenticity, so I had to wait. I appreciated their solidarity. Neither of them shrugged or said, if he doesn't make the sale, that's none of my business.*

By the example in the above story, you can't infer that the culture in Colombia is truly solidary. However, solidarity is a core value of our culture. So much so that we sorely complain when we see no displays of it. But cooperation, empathy, and compassion exist everywhere.

Thousands of simple acts like the above are commonplace, especially in the countryside. They're not the exception but the rule, although sometimes feel it's also in danger of extinction.

One of the most rewarding experiences of my job as a court-approved volunteer for the Guardian ad Litem (GAL) program in Florida, is that we act as a "next friend of the child" in court (not as an attorney). We advocate for at-risk children who've been separated from their families. And this organization operates on a system based on solidarity. Unlike in previous jobs or volunteer experiences I've had, in this organization I perceive that team members are authentically on the side of the child. And since they're on the child's side, they support their volunteers. They take our recommendations seriously. They don't look suspiciously at our actions. They don't assume that we might have secret agendas. As I said, the very essence of the organization is solidarity. Its motto is, "I am for the child."

In an ordinary job, the first and second responses to stress are easily activated. The first, especially when we feel our job is threatened, the second when we feel motivated to achieve and compete. But in a team, such as the GAL's, the crisis of a child who's been separated from his family has to be approached by activating the third, affiliative, response, which facilitates empathy, care and cooperation.

When altruism extends from the personal to the group or intergroup level, we're talking about solidarity, which is characterized by interdependency, mutuality, and fraternity.

There are movements throughout the world that seek to advance in the principles of solidarity based on cooperation, equality (race, ethnicity, gender, class), democracy, sustainability, pluralism, and above all, in an order of priorities where people and the planet come before the profiting of the few.

Solidarity is seen happening at different levels, the first of which, the individual level, manifests as the desire to help those perceived as equals in at least one aspect. People with whom one is associated either by destination (family, country) or any other circumstance. Often, solidarity is expressed toward the most vulnerable person in a group. It's easy to recognize and feel empathy for another human being when one considers the idea that we could very well be the one who needs help.

In Naples, the city where I live, traffic becomes slow and heavy in the first months of the year, when thousands of people come to find warmth in Florida, fleeing the winter cold in the north. We permanent residents, commuting to work or doing errands, get impatient. What's wrong with them? Why so slow? Most of the visitors are aged and retired and drive very slowly, either because they don't know the area very well yet, they have no rush to get anywhere, or they've lost their reflexes, their sight, or their hearing, so they drive with extreme caution.

It requires a little determination to be mindful. We have to make an effort to put ourselves in the shoes of the other and to also understand that in a few years our own sight may not be so sharp, our reflexes will slow down, and we'll be the ones driving with extreme caution and being the target of the impatience of others.

If we can't see ourselves in the situation of the other, our empathy will not be expressed. How can I understand the vulnerability of a gambling addict, for example, if I never liked gambling, if I judge gamblers?

At a second level, solidarity manifests itself in a group and can be induced by the behavior of the other members and sanctioned by the approval of the group in general. This is the case of support groups (Alcoholics Anonymous, for example) in which this type of fellowship is presupposed. It's also the case for a community affected by a disaster, where the good of all depends on the good of the individuals.

At a third level, solidarity manifests itself not only at the level of social norms but at the contractual level, as is the case in social welfare systems that include contracts between the state and other entities to provide services to the population. That would be the case of social security or compensation funds.

## Solidarity networks

Those who have a good support system are more likely to successfully survive a traumatic event. There's a significant correlation between psychosocial well-being after a trauma or an emergency and the solidarity and support received. The psychologist and researcher Peter Levine, author of *Waking the Tiger: Healing Trauma* (North Atlantic, 1997), and teacher of somatic therapies, has devoted his life to studying the physiological origins of trauma and has designed a therapeutic approach (Somatic Experiencing) to treat its aftermath. Levine draws attention to the fact that unlike humans, wild animals, living in daily situations of danger, don't seem to get traumatized or develop adverse symptoms, because "animals in the wild instinctively discharge all their compressed energy" and soon resume their normal functioning.

The memory of a traumatic event can't be erased from our minds; the synaptic connections resulting from the traumatic experience also remain. But what science teaches us today about neuroplasticity, neural circuits, gene expression, and behavior is encouraging. In brief, the brain can modify its structure and function throughout life, to respond to environmental changes, which also has an influence on molecular structures, gene expression, and behavior.

Levine explores the interconnections between evolutionary neurophysiology, animal behavior, and trauma in his book *In an Unspoken Voice: How the Body Releases Trauma and Restores Goodness* (North Atlantic Books, 2010). He points out that in the event of an accident (or a natural disaster, war, or surely a broken heart), it's essential for the victims to receive proper support immediately after and then to have the opportunity to rebuild medium-term social networks to restore a sense of belonging to the community. People who were already marginalized in the community before a catastrophe are more vulnerable and may be more likely to show mental symptoms after the traumatic event.

Social protection is attained by rebuilding support systems and activating community mechanisms that reduce risks and meet immediate needs.

Many circumstances make it clear that human networks play a key role in crises and disasters. Research supports the idea that life stress can be mitigated with the support of other caring people.

Isolation, sadness, loss of interest, attention deficit, sleep disorders, separation anxiety, and fear of death are among the emotional and mental reactions observed in subjects after they're exposed to traumatic incidents. Young people who feel isolated and are dealing with trauma can also display aggressive behaviors, which can further isolate them from their peers.

In the past two decades, several studies have come to the same conclusion: satisfactory group relationships and supportive relationships with family members or other adults protect people (especially young people) who've been victims of a traumatic event. Of course, there are also individual factors influencing the development of resilience (ability to recover after trauma). Social support consists of offering information, material help, advice, encouragement, and empathy to people in distress.

I recall a lecture by Colombian mathematician Germán Zabala[148] that I attended while I was in medical school. He had done his numbers and concluded that, from a strictly statistical point of view, the people living in the shanty towns up in the outskirts of Bogota should already have disappeared from the map. In the sixties, those barrios lacked

[148] The late Germán Zabala, was a mathematician at the Universidad Nacional and a co-founder of the universities INCCA and América in *Colombia.*

elemental health services, roads, and utilities. Many people lived in dwellings they'd constructed themselves from perishable materials (carboard, plastic, tin) in an area with a constant threat of mudslide following the rains. It was also a population of very limited economic resources, and in most cases, they couldn't procure adequate nutrition for their children. If they survived against all odds, Zabala assured, it was thanks to the existence of solidary systems based on informal credit, bartering, or trading of goods and services. It was a community that subsisted only because of mutual support. Many impoverished communities around the world survive because everyone has something to offer and individuals are motivated to reciprocate.

Historically, more than 98 percent of human evolution happened during the time our ancestors were still hunters and gatherers. According to what anthropologists know today, our ancestors lived in either cooperative tribes or bands. The entire tribe was responsible for the upbringing of children, and both sexes were involved in obtaining food. There was a division of labor: most men hunted, and most women were gatherers. Cooperation with the other members of the tribe or band was a priority, and this cooperation was essential to deal with climatic factors, confrontations with other tribes, and access to food.[149]

With the knowledge and tools we have today, cooperation, a sense of community, and bonds between human beings capable of watching out for each other could be the key to saving the planet. Solidarity requires taking responsibility, and the mutual dependence of people who feel obligated to each other. An important axiom of a solidary culture is egalitarian social relations. One's issues are of concern to all of the members of the group. It's acknowledging that our destinies are intimately connected. Unlike altruism, which leads to helping others through a simple moral commitment without the need for reciprocity, in a solidary community there is a form of cooperation in which everyone works for the common good.

Traditionally, unions, some nonprofit organizations, community and civic associations, community-based childcare and volunteer groups, cooperatives, and some political parties are guided by philosophies promoting solidary behaviors among their members or by their members toward certain sectors of the community.

---

[149] J. A. Simpson and L. Beckes, "Evolutionary Perspectives on Prosocial Behavior: The Better Angels of Our Nature." In M. Mikulincer and P. R. Shaver (Eds.), *Prosocial Motives, Emotions, and Behavior: The Better Angels of Our Nature* (pp. 35–53) (Washington, DC: American Psychological Association, 2010).

# PART EIGHT: The Love Possible

*The heart of man is like a musical instrument, it contains great music. Dormant, but it is there, waiting for the right moment to be interpreted, expressed, sung, danced. And it is through love that the moment arrives.*

—Rumi

In her TED talk about revolutionary love, civil rights activist Valarie Kaur exhibits her wisdom, compassion, and strength. She exudes poetry when, looking at the state of the nation, she says to us, "What if this is not the darkness of the tomb but the darkness of the womb. *Is our nation dead or still waiting to be born?"*

She's asking us to reclaim love as a revolutionary act. I invite the reader to explore her Revolutionary Love project further.

We began this book by stating that we're hardwired for empathy, compassion, and love. Then, we critiqued the current state of affairs in the world and discussed some serious obstacles along the path of love. Now comes the time to explore how to move forward.

It's not enough to be physiologically equipped for love. It's not enough to be aware that, from the evolutionary point of view, love constitutes an adaptive advantage. The key question appears to be this: if we're fitted for love and it's an evolutionary advantage, why does it seem so hard for us to love each other, to love ourselves?

Many evolutionists support the idea that empathy and compassion have become deep-rooted traits in human beings because they contribute to the survival of the species. Darwin himself predicted that the human capacity to feel empathy was an adaptive advantage that would be extended by natural selection. As already mentioned, he also called it our "strongest instinct."

Neurobiology has identified neural circuits that intervene in the regulation of our emotions, in the way we respond to stress, in the way we relate to each other. In particular, the activation of an affiliative tend-and-befriend system seems to be involved in the formation of the bonds we establish between us. We've also discussed our innate need to connect with others and some of the factors that maintain a separation mindset.

In trying to answer why, if we're biologically equipped for love, we feel often separated, we examined obstacles standing in the way of love but stated that even these interpersonal hurdles are the result of how societies evolve amid the contradictions between those who own the means of production (land, capital, corporations) and those who own nothing else than their capacity to work. Now, we visualize that substantial changes are forthcoming in an overpopulated world, where water and other vital resources we depend on are already becoming scarce, while senseless industrial greenhouse emissions and consumerist lifestyles continue to contribute to accelerate climate change. Too much is at stake.

The next question is then, how can we contribute to confronting these challenges and finding solutions for them?

## Evolving consciously

*Life has no choice but to accept the challenges imposed on it at every step by its multiple relationships with the outside world, no matter how fleeting they are or appear to be; the necessary ends up being modified by necessity. Chance ends causality. The exception ends the rule. Fortuitous circumstances end the law. That is the key to everything.*

—Francisco Mosquera

A majority of us recognize *Homo sapiens* as the product of the natural evolution of species. Institutions and systems (marriage, government, the economy) also evolve over time. Human beings and their endeavors are fluid, not static. It's easy to grasp the concept that our minds, our mores, our identities, our political ideas change thanks to the contradictions and crises that drive our growth.

At all times we're making choices, albeit very often unconsciously, between several options. But just imagine what would happen if most of our choices, or at least the more crucial choices, were conscious and driven by the common good? What if we could grasp the extent of the impact of what we choose to be, do, say, own, and think? What if we could imagine the consequences our choices might have in our own destinies as individuals and our destiny as a species? What if we could clearly envision our impact on nature, institutions, and relationships?

What if we consciously decided that fostering empathy, compassion, and solidarity needs to be at the core of our parenting and education models? If we would start valuing people not based on their intellect, power, appearance, or status but on their capacity to be compassionate.

Paul Gilbert tells us that a compassionate focus is knowledge based. Empathy needs to be cultivated. Gilbert cites Harvard professor emeritus Edward O. Wilson: "The more closely we identify ourselves with the rest of life, the more quickly we will be able to discover the sources of human sensibility and acquire the knowledge on which an enduring ethic, a sense of preferred direction, can be built."

We must begin by understanding in which ways our individual destinies, the nature of our institutions, the interests of corporations, and the fate of the planet are interconnected.

Chaos theory, as it's popularly known, is a branch of science that deals with the unpredictable. It studies certain types of complex and dynamic systems, that are very sensitive to minute variations in initial conditions. It uses the metaphor known as the "butterfly effect" to explain the existing interconnection among events in the universe and it's formulated like this: The fluttering of the wings of a butterfly in Hong Kong could cause turbulence that might eventually become a storm in the Caribbean. In practical terms, this means that any small variation we introduce into a system can lead to random, unpredictable events. It helps us think that contributing a tiny grain of sand could actually, somehow, influence our collective destiny.

Let's add Erwin Schrödinger's entanglement theory to the intricacy of this subject, which deals with the mysterious atomic particles that remain interconnected in such a way that the actions performed on one affect the other, even at great distances. This should help us further understand how intimately linked everything is in the universe.

Briefly consider what it means to stop being impotent, passive, subjects of the state of affairs in the world and instead, knowing with certainty that we have the ability to consciously influence the course of life, the health of the planet, the future of humanity, the evolution of the world. Wouldn't this certainty compel us to be more responsible for our decisions and actions?

Let's consider that if ignorance of the law excuses no one, neither ignorance of moral obligations excuses us.

The idea that as our consciousness evolves, we are progressively in charge of our destiny could be framed within the theory of conscious evolution, which proposes that the human species, for the first time in recorded history, is in a privileged position of controlling its destination, based on the available knowledge of the current challenges we face. According to the "Evolutionary Manifesto" (available online at www.evolutionarymanifesto.com), "a completely new phase in the evolution of life on Earth has begun. It will change everything. In this new phase, evolution will be driven intentionally, by humanity."

Science, technology, and spirituality (understood as the awareness of our interconnectedness) provide the tools for this conscious change.

Paul Gilbert points out that in recent years, our understanding of compassion has deepened. Medical personnel and psychotherapists have begun to recognize empathy and compassion as antidotes to conditions such as depression and anxiety. There is increasing evidence, Gilbert says in his book *Overcoming Depression* (Robinson, 2009), that learning and practicing compassion can change our brains. Citing Buddha's teachings, Gilbert advocates for the development of insight into our thoughts and passions (mindfulness). Also, to cultivate compassion as a major remedy to suffering and unkindness to self and others. Since the mind is unruly and, without training, run by primitive passions, our lives are sometimes governed by the pursuit of short-term pleasures and efforts to escape and avoid disappointments, pain, and suffering. As we've discussed, compassion stems from consciously nurturing the ability to display an empathic response.

What turn would human evolution take if we collect the wisdom acquired through personal experience and transmit it to the next generations in a systematic and conscious way? I can't concur with the individualistic affirmation, often applied to the upbringing of children, that each one should learn from their own experiences and mistakes. For me this assertion is the equivalent to telling school children that they have to rediscover the paramecium or the formula for the water molecule or the value of the hypotenuse. Or worse, that they could take any risk they want, testing a behavior by themselves to see if it's dangerous.

Perhaps we need to resume the systematic oral transmission of wisdom gathered by parents and elders, through stories and fables, myths and epic narratives, proverbs, and songs focused on different aspects and expressions of love.

And consider this definition of wisdom offered by psychologist Ronald Siegel at a lecture in Florida: "Wisdom refers to our concern for the effect our actions have on others." And ourselves, I would add.

Transmitting this wisdom would also require a conscious evaluation of the validity and benefit of predominant traditions, behavioral patterns, and beliefs to fluidly adapt to the demands of the present time.

What would we do differently if most of the time we could hold this awareness of being responsible for our destiny? What new movements would we create? How would we relate to other sentient beings? What would we do to transform the world into a kinder, healthier place?

What decisions and actions would we choose if we willfully asked ourselves how to be more loving, more compassionate? If we advanced through the different stages of life with the intention of consciously mastering the different aspects of love? If we asked ourselves, "How could I love my fellowman and the planet in a more effective and positive way?"

How would it be if instead of isolating and excluding the "trespassers" (and those who harmed the community by not behaving in a loving manner), we found a way of helping them remember who they truly are.

Alice Walker[150] tells a story of the Babemba tribe of South Africa:

> When a person acts irresponsibly or unjustly, he is placed in the center of the village, alone and unfettered.
>
> All work ceases, and every man, woman, and child in the village gathers in a large circle around the accused individual. Then each person in the tribe speaks to the accused, one at a time, about all the good things the person in the middle of the circle has done in his lifetime. Every incident, every experience that can be recalled with any detail and accuracy is recounted. All his positive attributes, good deeds, strengths, and kindnesses are recited carefully and at length.
>
> The tribal ceremony often lasts several days. At the end, the tribal circle is broken, a joyous celebration takes place, and the person is symbolically and literally welcomed back into the tribe.
>
> The rule is that no one should fabricate or exaggerate the person's attainments or good virtues.

This beautiful story is an uplifting example of a society that opts for integrating instead of excluding trespassers.

## Overcoming fear, making room for love

Although fear and suffering are inexorable parts of our lives, they should not be our masters. Responses to dread and pain are learned and memorized, and they can also be regulated. Buddhists use meditation, fasting, and contemplation to facilitate stepping out of the domain of suffering, which is where the mind becomes reactive to physical or emotional pain. This requires strong willpower and practicing a healthy detachment from mundane affairs.

[150] A. Walker, *Sent by Earth: A Message from the Grandmother Spirit after the Bombing of the World Trade Center and Pentagon* (New York: Seven Stories Press, 2001).

Here are some strategies that could help us with the regulation of emotions and the development of empathy:

1. Be kind to your body: a healthy lifestyle includes awareness of how your emotions affect your physiology. Eat healthy, exercise, learn to manage stress levels.

2. Learn from the acts of kindness of others, their sacrifices, and integrity. Be reminded of all the potential we have as human beings.

3. Work with determination on getting to know each other better. In many cases we're not perceiving what is, and instead we use a mental mechanism called projection. If you find yourself being judgmental and that judgement is becoming an obstacle to connecting with others, ask yourself: Might I be seeing in others those aspects that I dislike in myself? Am I replacing missing information about the other with assumptions? Projection will not allow room for empathy.

4. Recognize our common humanity. It helps us stop judgement (of self and others). It requires listening carefully and clarifying any misunderstandings promptly.

5. Take responsibility for our actions and for our emotions instead of blaming others for what we felt or did.

6. Reinforce emotional regulation processes through the practice of active listening and being present. Learn to manage stress by practicing, for example, breathing and relaxation techniques, meditation, Reiki, Qi Gong, and mindfulness. Remember to look for those "micro-moments of connection" to stimulate the vagus nerve and slow down your heart.

7. Therapeutic interventions that use mindfulness, loving kindness, and compassion are now widespread in the United States and elsewhere. Mindfulness-Based Cognitive Therapy (MBCT), Dialectical Behavioral Therapy (DBT), and Acceptance and Commitment Therapy (ACT) are designed to direct the person to focus on the present experience without judgment.

8. Find tutors, mentors, guides, counselors, or companions to support the process of knowing yourself.

9. Foster the creation of parenting circles (including teachers and therapists) in the community and in schools to promote a type of parenting and education based on love and not on fear.

10. Create a practice that facilitates cultivating compassion for self and others.

Learning and practicing love and empathy consciously and using our learning process for educating others could help us in transforming our lives and our environment.

## Multiply your acts of kindness

*The City: Portland Oregon, in the northwest of the United States. The scene takes place on a train. A young man, thirty-five years old, later identified as Jeremy Joseph Christian, enters the train, crosses the car, and grabs a pole near the opposite door. He starts shouting: "I pay my taxes. I have rights, and people of color are ruining my city. I have the right to freedom of expression, right that the constitution grants me."*

*Then he yells obscenities and says Muslims must die. His words, like darts, are shot directly at the faces of two young girls, one with black skin, the other with the face and head half-covered by a hijab.*

*The outcome of this expression of hatred? Two men are killed with a knife, and a third one survives the wounds after trying to calm the man driven by anger.*

*The families of the victims, their friends, and other residents of Portland later meet, embrace, say supportive words, express their admiration for the act of courage displayed by the men who protected the girls, saving their lives.*

*One of the victims, Taliesin Myrddin Namkai Meche, only twenty-three years old, will be remembered as a person full of compassion. He had big plans. He was starting his career at an environmental consulting agency. He had just bought a house, and he dreamed of getting married and having a family. He died protecting the two Muslim teenagers attacked by Christian. The other victim was Rick Best, fifty-four, a retired military officer who worked as an electronics technician. He lived with his wife and four teenage children in the suburbs of Portland. Three years before, he had been a candidate for public office, inspired by his vocation of service. At the time of the attack, he worked for the city.*

*A third men, Mika Fletcher, twenty-one, suffered serious injuries but survived Christian 's attack—against three unarmed men— with a knife aimed with precision at their jugular veins. (Story built from various media sources)*

Those who suffered physical injuries were not the only altruistic souls. Several passengers on the train stayed to offer help to the injured and waited until the ambulances arrived. All of them will remember the horror and wonder how a human being could behave as Jeremy Christian did.

Two opposite kinds of human beings were found here, showing themselves at the same time. But that's where our hope comes from. We are capable of the worst but also of the best, and we have choices. This certainty is key in a stage of great uncertainty.

Acts of kindness and sacrifice inspire us, demonstrate our potential.

## Educating in love

*Loving without knowing how to love wounds the person we love. To know how to love someone, we need to understand them. To understand them, we have to listen.*

—Thich Nhat Hanh.

At the age of nine, a moment of inspiration set me on a path to becoming an educator. I felt, rather than knew, that something was wrong with education, both at home and at school. I would have to reread *Little Men* (Roberts Brothers, 1871), by Louisa May Alcott, to fully understand the impact this book had on me at that early age. The Plumfield Estate School, run by Mama Baher (the protagonist) with her husband, seemed like a paradise where children were treated with respect, empathy, and affection, but, above all, where there was an understanding that the students were children, and teachers allowed them to be so.

Although I had already been born with a call to heal (a desire to become a doctor that goes as far back as I can recall) and had also made my first attempts at writing stories, I promised to myself that one day I would create a different kind of school. It took me thirty-four years to gather enough faith in myself to dare to start one.

The school project came together thanks to the selfless dedication of the members of our Fundación para el Desarrollo del Joven—fundeijoven—created in 1991, and later thanks to the support of a relative, Margot de Pellegrino, founder of the *Fundación para la Actualización de la Educación* (FACE), in Bogotá. Without the background of my youth work in Magangué in 1986, a youth project developed in Barrio Chiquinquirá in Cartagena, between 1988 and 1989, and the successful experience of our Carpe Diem school in Cartagena, I would not claim any authority to talk about educating in love. Ours was a very fruitful experience. In our model of education (1991 to 2001), we created an environment with zero tolerance of any form of violence, and children were never coerced into studying out of fear of not passing tests or of getting failing grades. Students evaluated themselves according to the objectives they had previously set for themselves. Teachers learned to avoid labeling children, understanding the negative impact tags can have on the formation of a child's identity.

We were further encouraged by Summerhill School, an independent British boarding school founded in 1921 by Alexander Sutherland Neill, who believed schools needed to fit the child's needs and talents instead of imposing standard curricula that disregard we're all different. They understood play as an invaluable learning opportunity.

At Carpe Diem, we believed that a student-centered education that respects the individual's pace and interests renders positive outcomes. Children's questions were encouraged, and they learned to formulate hypotheses and search for answers. It stimulated their critical thinking and mathematical skills. Children had direct access to books in the classrooms (use of computers was still limited).

British academic and learning innovations consultant, Steven Wheeler says, "True pedagogy is far more that someone instructing. Pedagogy is leading people to a place where they can learn for themselves." But even in a digital era, it doesn't work that way. In many cases, education fosters dependency.

The learning process we implemented at Carpe Diem allowed children to access information sources, they became skilled in processing data, and then they found ways to apply what they learned to real life. It facilitated the acquisition of advanced cognitive abilities.

Our pedagogical innovation was intent on eliminating fear. It aimed at becoming a model for educating in love. No doubt fear is a strong motivator. It leads you to do whatever prevents pain. It works as an external emotional regulator. But educating in love involves applying empathy in the educational process. It leads to self-soothing, to introspection, to the strengthening of bonds, to self-motivated learning, and to joy.

Carried out on a small scale, our experiment was an exception to the rule in Cartagena. But the school FACE founded in Bogotá, continues to grow thirty-five years later, demonstrating that educating in love is viable. There are FACE schools in the Colombian cities of Bogotá, Cúcuta, Armenia, Santa Marta, and Villavicencio. And many other schools have adopted FACE's philosophy.

Some children came to Carpe Diem after experiencing difficulties in other education centers. Several had lost motivation for learning and felt frustrated at not achieving what was expected of them in their previous schools. Some had trouble socializing, which seemed related to highly competitive environments where they had experienced bullying. But often, it was the unrealistic expectations of their parents and teachers—frustrated by certain behaviors or because the children were falling short of high standards—that seemed to be at the root of the children's behavior.

In our experience, an institution concerned more with academics or achievement than with a student's process runs the risk of neglecting the emotional development of the child in all its different components: affect, safety, the ability to be assertive, socialization, and the management of sexuality.

Besides the family, the school is one of the most crucial social factors influencing the emotional maturation of the child; therefore, it's also decisive in the development of cognitive processes (attention, memory, perception, and observation). But schools can also have a significant impact on the emotional and social development of the child. Therefore, they must aim at creating anxiety-free environments while contributing to nurturing and gratifying the emotional needs of the child, promoting curiosity, allowing exploration, and stimulating mastery of certain skills and talents.

The spiritual leader Osho said that schools should focus on teaching the art of living, the art of dying, and meditation (in addition to some English, science, and mathematics): "A real education will not teach you how to compete; it will teach you to cooperate. It will not teach you to fight and come first. It will teach you to be creative, to be loving, to be blissful, without any comparison with the other. It will not teach you that you can be happy only when you are the first."

Do schools teach children how to love others? Do schools show children the best ways to love and respect their bodies? What about children learning about the responsibility we all have in the preservation of the earth?

It's sad to see how many children eat plenty of foods laden with empty-calories or fats (junk food) or subject their bodies to exercises and fashion regimes without grasping the long-term negative impact those might have on their bodies. At Carpe Diem, children learned they had choices, but eating junk food was not one of them because, why would you put harmful things in your body? To make a healthy decision you need to first fully develop your awareness. Teachers (and parents) need to direct the attention of the child to what is best for them.

The issue of loving the body deserves special mention in a consumer society that seems to promote a progressive and slow murdering of our bodies. I'd dispute the idea that allowing children to choose certain foods is love. Why feed our children with foods that lead to chronic inflammation and illnesses? Only because we're not making healthy choices ourselves. Please note the terms we use reflect the treatment we give our bodies: you *kill* yourself working; you eat *crap*; you party till you *drop*; you might compete *to death.* These expressions are woven into the fabric of a culture that steadily disrespects the body. This is not to mention how widespread the abuse of mood-altering substances and pharmaceuticals has become.

The benefits of learning to love extend beyond oneself, but they begin with self-knowledge and the development of self-compassion.

In the *Art of Loving*, Fromm explains how the practice of any art—including the art of loving—requires discipline, concentration, patience and dedication, without which no art can't be mastered.

In our school, Carpe Diem, our schedule offered two weekly hours for self-exploration. It was a safe space for the children to examine relational issues and learn to express their feelings openly. "Safe space" refers to a place and moment in which a person can feel comfortable and sheltered. Where they can express themselves freely and gain insight, knowing that they're listened to and accepted, and that what's said is confidential. Once a safe space is created, it becomes easier to express emotions and develop a healthy capacity to regulate them. In these group sessions the students were able to put on the table any grievances or conflicts existing between them or between them and their teachers, and this gave them the opportunity to mature ways of solving conflict. Sometimes they watched a movie and discussed its content or examined their lifestyle and its impact on their bodies, on others, or on the world. They used music, painting, drama, or body movement to express themselves. It was a time, in short, for reflection and introspection.

The years have proven that our methodology had a positive impact on the lives of the children we served. The results reaffirm the idea that compassion can be taught, love can be learned, and fear can be excluded from education. Also, that we can offer models of solidary relationships and teach principles of cooperation.

It seems natural for children to respond lovingly. However, it's important to invite them to look at the different ways in which others experience the world, helping them to reflect on the impact their actions have on others, on the planet, and on their bodies.

Much has been said about bullying. One of the ways to prevent harassment implemented at Carpe Diem included a very simple activity. When a student with special needs was admitted, we invited other students in the group to talk about their own challenges. When classmates reflected on their own needs or limitations, the new child relaxed. We embraced our common humanity, acknowledging that we all have some limitations that prevent us from functioning fully. These might be laziness, obsessing, or limping. Someone might have a stiff knee or headaches. Others might suffer from dyslexia or a visual deficit. Some children have to deal with extreme shyness, others with social anxiety. Part of our life journey has to be precisely about dealing with or overcoming our limitations.

According to research published in 2014 by the British not-for-profit organization Scope, about 67 percent of Brits felt uncomfortable when talking to a disabled person. A survey by Louis Harris and associates had found similar results in the US in the nineties. However, we can become aware of, and relate to the discomfort, if we use cognitive empathy to try to understand what another person is feeling.

When a child is going through emotional turmoil, one of the most common reactions from peers is to turn away (flee) because they don't know how to handle the stress the situation elicits. We invited the children in Carpe Diem to open up and try to understand what the other child was going through and then to think of what they could do to support their peer. Most children responded positively to our suggestions. Being supportive is natural.

Stimulating empathy in children is one of the key objectives of inductive discipline. In this type of discipline, social transgressions are not approached with punishment.

Most modern educators are aware that punishment for social transgressions engenders reactions ranging from resentment to defiant behaviors. Instead, a child could be induced to feel sorry for the discomfort he might have caused and helped to reflect on the effect his actions had on another. Then a reparative action can be suggested—hugging, asking for forgiveness, inviting the other to play—so that shame and guilt are attenuated. These behaviors would be remembered and would eventually contribute to a reinforcement of the neural circuits for empathy. The newfound empathy will then contribute to limiting aggression and increasing prosocial behavior.

In *Unselfie: Why Empathetic Kids Succeed in Our All-about-Me World* (Simon & Schuster, 2016), psychiatrist Michele Borba expressed why we want kids to empathize:

> For starters, the ability to empathize affects our kids' future health, wealth, authentic happiness, relationship satisfaction, and ability to bounce back from adversity. It promotes kindness, prosocial behaviors, and moral courage, and it is an effective antidote to bullying, aggression, prejudice, and racism. Empathy is also a positive predictor of children's reading and math test scores and critical thinking skills, prepares kids for the global world, and gives them a job market boost.

A parent group at the school or in the community is an ideal space for them to learn about empathy. They can start by sharing their experiences about their upbringing and the circumstances in which they grew up. Then they can identify their beliefs about discipline and explore new parenting models that include listening, accepting and understanding.

Autocratic or authoritarian models are not fertile grounds for love. Most parents and teachers raised in authoritarian systems will tend to discipline children using punishment and reward, as they learned at home or in school. Others will become permissive, trying to avoid repeating what was done to them. In a hierarchical, vertical plane, authority is imposed, and there is no participation in the exploration of needs, conflicts, or solutions or in the building of consensus. The result is dependence, lack of autonomy, fear (if not defiance), and opposition. The responsibility lies entirely with those who hold authority, and children do not participate in the decisions affecting them.

Instead, on a horizontal plane of movement (see pgs. 41-42), children feel capable (they're allowed to do), feel included (they're allowed to participate), and know that they count in the family or classroom (that they're active subjects, not passive ones).

## Nurturing children for wholeness

Swiss psychoanalyst Alice Miller used the term "poisonous pedagogy"[151] to criticize a series of parenting practices typical of families and schools where an authoritarian model prevails. Children are repressed, punished, and abused, and then told that it's for their own good in an attempt to validate the abusive punishment, contending that the child asked for it. This kind of discipline, its practitioners argue, builds character. Miller explains: "Poisonous pedagogy is a phrase I use to refer to the kind of parenting and education aimed at breaking a child's will and making that child into an obedient subject by means of overt or covert coercion, manipulation, and emotional blackmail. The mistreated and neglected child is completely alone in the darkness of confusion and fear."

We need new parenting strategies based on compassion, which is the foundation on which solidarity is built. No, it isn't enough to feel we love our children with all our hearts or to fulfill their material needs. Our deeds

---

[151] Miller has written several books worth reading. I especially recommend: "The Drama of the Gifted Child: the Search for the True Self." *(New York: Basic Books, 1996)* and "Thou Shalt not be Aware: Society's Betrayal of the Child." *New York: Farrar, Straus and Grouts, 1981.* She discusses "poisonous pedagogy" in her blog: www.alice-miller.com.

(not just words) also need to speak of how we love one another, and I think here is where we fail as parents, educators, and society.

I guess I will never stop emphasizing that the solution to the problems we experience in the world rests on how we human beings are raised to develop our capacity for empathy and compassion. The road to peace and a better understanding among human beings will never be built on power struggles, forcefulness, and threats.

A mother who constantly screams and loses control, a father who slams doors and beats or threatens a child, can't expect their three-year-old child will grow to be a sweet, calm and obedient person. If anything, he or she could turn submissive, but more likely, subversive (not a bad thing if it's about transforming a corrupt, violent, or inefficient order).

Physical punishment between ages six and nine predicts higher levels of antisocial behavior, starting just two years later.[152] On the contrary, there is no evidence that physical punishment works as a parenting strategy. We need to start adopting new parenting styles, new pedagogical approaches. Parents need all have access to programs that introduce positive parenting techniques, which have proven successful.

Martin Hoffman, cited above, believed that inductive discipline (unlike power-assertive discipline) can promote empathy and decrease guilt and shame in the child. In inductive discipline, the child is given the opportunity to put himself in the shoes of the other and display a reparative behavior to compensate for any wrongdoing.

Krevans and Gibbs[153] corroborated Hoffman's theory. Children raised with inductive discipline were more empathic. They also found that contrary to expectations, parents' statements of disappointment also positively correlated with children's prosocial behavior.

Inductive discipline, associated with democratic parenting, is characterized by affectivity, clear communication, and active listening. It takes into account the emotional needs of children and their interests but may include a high level of demands and control. The desired outcome is that this type of upbringing would contribute to autonomy and self-control and to the development of self-esteem and responsibility.

---

[152] J. Durrant and R. Enson, "Physical Punishment of Children: Lessons from 20 Years of Research," *CMAJ* 184 (12) (September 4, 2012): 1373–1377.

[153] J. Krevans and J. C. Gibbs, "Parents' Use of Inductive Discipline: Relations to Children's Empathy and Prosocial Behavior, *Child Development* 67(6) (December 1996): 3263–77.

In order to promote prosocial behavior in the children, it is necessary to build safe family and school environments, with a low level of conflict, where empathy can flourish.

Thanks to the pen of great writers like Charles Dickens, who portrayed the mistreatment of children in schools during the Victorian era, we have a clear sense of what should not be. But fast forward almost two centuries and we still find that psychological aggression (shouting, screaming at a child, calling a child offensive names) and corporal punishment (shaking, hitting, slapping, spanking) are still common.

The frequency of the use of psychological aggression and physical punishment as parenting methods is periodically evaluated by UNICEF. In 2017 about three hundred million children ages two to four around the world had experienced violent discipline by their caregivers on a regular basis. These parenting methods deeply affect young children. "Regardless of the type, all forms are violations of children's rights" (UNICEF).

Parents and educators continue to focus on modifying children's behavior, even if it takes to use physical punishment and psychological aggression. We need to ask what's the psychological impact of time-outs, ostracizing, or taking away privileges. Brain research suggests that isolating children from dear ones causes relational pain. Daniel Siegel explains that relational pain uses the same neural paths in our brain as physical pain.

Unfortunately, even psychotherapists often focus on behavioral interventions. Have they worked? Depends on what results you're looking for: controlling behavior, prompting compliance or fostering self-reliance, confidence, and prosocial behavior?

There are still countries condoning particular forms of physical punishment that in other places are considered abuse. There is strong evidence that pain (both emotional and corporal) inflicted by a caregiver disrupts both parent-child attachment (causing what is known as reactive attachment disorder) and the child's emotion regulation.

So how can we shift our focus from the child's behavior to nurturing the child's emotional needs?

Alfred Adler, a colleague and later critic of Freud, was interested in the child's relationship with his parents and how it affected their behavior. Adler believed that the central motivation of all humans was to belong and be accepted by others.

Influenced by Adler's writings, Rudolph Dreikurs, came up with a model of social discipline. Assuming that all children were looking to fit in, but that their assessment of the environment was subjective, he concluded that misbehavior was the result of a child's wrong guess about how to fit in and gain some status. Dreikurs identified four goals that motivated the child's misbehavior: attention-getting, obtaining power and control, exacting revenge, and displaying helplessness or inadequacy. He believed a misbehaving child was a discouraged one, and therefore he proposed the use of encouragement (not praise).

Dreikurs proposed a parenting and pedagogical method based on understanding the purpose of behavior and stimulating cooperative behavior. He believed all children are motivated to grow and develop and that we would find the unconscious purpose of their behavior if we asked what need is met by the behavior and what happens as a result of the behavior. Adler's school, based on Dreikurs's method, aims at graduating professionals that could contribute to alleviate social and global affairs.

Both Dreikurs and Adler referred to the parenting style they advocated as democratic. A democratic parenting style would allow the child to feel loved and accepted, confident of being able to overcome difficulties, aware of his achievements and contributions, but also aware of improvements that still need to be made. This child would see the world as a safe and friendly place without being afraid of making mistakes. In the democratic parenting style that Adlerians propose, children are seen as an integral part of the family. They're allowed to cooperate and to contribute their share. In Adlerian schools, children are offered appropriate challenges for their ages, in a way similar to that found in Montessori schools. Children are also allowed to advance and learn at their own pace. As mentioned before, Dr. Jane Nelsen built on Dreikurs ideas to develop "positive discipline" as a better alternative to a system based on punishment and reward.

Conscious evolution would include adopting pedagogical methods that effectively respond to the challenges presented by the current socioeconomic context. We need a holistic pedagogy that promotes harmonious coexistence.

In *Free Schools, Free People*, (Sunny Press, 2002), educator Ron Miller, a pioneer in holistic education, said, "Holistic education is a philosophy of education based on the premise that each person finds identity, meaning, and purpose in life through connections to the community, to the natural world, and to humanitarian values such as compassion and peace. Holistic education aims to call forth from people an intrinsic reverence for life and a passionate love of learning."

Neuroscience teaches us that our brain, very much like a garden, is alive and continually changing, not static but always adapting. We're still far from fully understanding how the new knowledge about the malleability and plasticity of the brain can be positively applied to the upbringing of children or to rewire adults' brains. However, the new scientific advances provide evidence that the world would benefit from developing pedagogies that stimulate a healthy brain development.

We can promote parenting in and for love, based on mutual respect between educators, parents, and children. Such a system would favor an understanding of the child's emotional needs of belonging and feeling significant.

We need to avoid creating educational and parenting models that, instead of promoting autonomy, critical thinking and creativity, focus on what we think the child should learn or achieve. Our impossible standards lead to forceful attempts at shaping children's behavior, emotions, and cognitive processes.

## Learning from the Semai

In certain communities there is a strong collectivism, as in Malaysia's Semai tribe,[154] which has a population of about fifty thousand. Disputes are resolved in public assemblies in which the reasons for a complaint are discussed, sometimes for days, so that everyone can offer their opinions. The discussion ends with an admonition by the head of the tribe, noting that conflict puts the community at risk. Experiencing shame in public serves the purpose of preventing the recurrence of unwanted behavior.

It's interesting that among the Semai, children's games are not competitive. Instead, they have the purpose of stimulating physical activity until the body gets tired and is ready to sleep and dream. If they use modern games, they modify them. For example, in badminton, they don't use a net and don't keep score. The shuttlecock is hit so that the blow can be answered by another participant. The goal is the game itself, not the win. Noncompetitive play stimulates cooperation. It shouldn't surprise anyone that they are a peaceful society.

We could learn some parenting strategies from communities like the Semai—very different from those prevalent in the West. They don't use a reward and punishment system. Their children are not forced to do what they don't want to. Instead, they're taught to fear the forces of nature but also how to control their own aggressive impulses. Children soon learn an important skill. They're taught to preserve peace among them by applying a simple method: give in! Yes, being right, enjoying

[154] Read more about the "Semai and Peaceful Societies" on the University of Alabama at Birmingham's site: https://bit.ly/2v8L1tc.

privacy, having their own space and possessions (having what one as an individual wants), is not as crucial as preserving harmony in the community. This stands in great contrast to the prevailing relationships in an individualistic world dominated by the intellect, where it's more important to be right than to be friends.

Communities like that of the Semai could help us to better understand the possibilities of contributing to building (rebuilding) a society in which the priority would be the common good, and not my individual freedom, my desires, my stuff, my accomplishments. One in which freedom lies precisely in understanding that my choices must take into consideration the common good.

On the other hand, the vindication of individual freedom (or individual sovereignty) seems to have factored into our disconnection from each other and the planet. It might explain why systems based on solidarity are disappearing, although they were part of ancient cultures and are still observed in most Third World countries. A society driven by greed can hardly be friendly.

A close look at the life of tribes like the Semai, which have managed to preserve a communal lifestyle, would also help us design parenting strategies to stimulate autonomy but avoid excluding the other. There is a difference between individual freedom (which potentially isolates and dissociates us) and autonomy (which allows us to make conscious decisions, weigh the consequences of our actions, and take the common good into account). Individualism, promoted as the ultimate expression of liberalism and freedom, might have advantages—most of all for a socioeconomic system based on consumerism—but freedom is very problematic without the other two components of what, at the time of the French Revolution, defined the dawn of a new era: equality and fraternity. Although the purpose of the French revolutionaries was to conquer human rights (freedom), it was also to use the political freedoms obtained by working together for social and economic justice. However, that freedom has taken an inevitable turn toward individualism, typical of a utilitarian society, which is the essence of capitalism.

## Brain-minded parenting

The good news is that we have a better understanding of how neuroscience (and mindfulness) can be applied to the process of parenting. For ages, we've been (mostly) mindlessly adopting parenting and pedagogical strategies that we assumed worked, without linking the ills of society to the ways we raise and educate children. Freud's main contribution, I think, was to open us to the understanding of how childhood experiences define the rest of our lives. Neuroscience is now taking us beyond that understanding to show us that even an adult brain can change, which means that, with proper stimulation, we can strengthen the circuits of empathy.

Exploring the science of parenting, I try to answer key questions:

How could neurobiology be applied to raise healthier, fulfilled, empathic, and socially skilled children?

What role does fear play in the development of the neural circuits involved in parenting?

How does parental stress affect the quality of parenting and the child?

How do early experiences affect the development of the child's brain?

How could we shift our focus from forcefully modifying the child's behavior to satisfying the child emotional needs?

Daniel A. Hughes and Jonathan Baylin's work provide key answers to the neurobiology of parenting. They make an important point: the hearts and minds of parents do affect the developing brains of children.

The authors also help us understand how stress could hijack parents' capacity to offer loving care. They identify five brain systems or clusters of neuronal circuits involved in good parenting:

1. **The parental approach system**, which enables parents to be close to children without becoming defensive. It's activated by positive interactions between parent and child and is mediated by dopamine and oxytocin. The activity of the amygdala is toned down.

2. **The parental reward system**, which makes it possible for the parent to enjoy the experience of parenting. The vagal system is engaged, facilitating the parent-child connection.

3. **The parental child reading system**, which supports the ability of a parent to understand and empathize with a child's inner subjective experiences. The parent's mirror neurons help him or her perceive, detect, and understand the child's reactions and emotions.

4. **The parental meaning-making system**, which allows the parent to construct a working narrative or story about being a parent, involving the upper brain.

5. **The parental executive system**, a neuronal circuit that relies upon the dorsolateral prefrontal cortex and helps the parent to regulate the lower, more automatic brain processes. It allows conflict resolution.

Parents need to learn to monitor their own feelings and actions. Mindless parenting leads to the repetition and transmission of patterns, habits, and ideas acquired in their family of origin or learned from society. It's important to understand that what we do as parents will continue to affect generations, until someone breaks the automatic reprise.

Hughes and Baylin explain how what they call the brain's executive system allows us to process information about the degree of attunement with our children. At times, parenting is unrewarding, and it can become overwhelming. In other instances, we just don't have the insight or the inner resources to respond to the child empathetically. In those moments engaging the executive system seems crucial to resolving inner conflict between parental and "unparental" feelings. We need to make time to reflect on our parenting experience. The parental executive system, heavily dependent on the functioning of the upper brain (neocortex), helps keep the other four systems turned on and working.

When we go through difficult times, such as existential or identity crises (intrapersonal conflict), or when there is conflict with family members, peers, or friends (interpersonal conflict), our parenting circuits might be affected. We need time for introspection, relaxation and "rebooting" so that stress doesn't interfere with our caregiving. Easier said than done, right?

And what if we were not blessed with good parenting skills? If our parents were abusive or neglectful or just "not there?" What if we never acquired the skills necessary to respond to the challenges presented by our interactions with a child? How do we stay engaged, calm, and mindful as parents? Counseling is, of course, a good resource.

But I see hope and strongly advocate for the creation of "circles of parents" in schools and in the community, with the participation of educators and therapists. In these circles, self-exploration would be a very valuable tool. The facilitators would help participants understand how their upbringing shaped their minds and hearts and how it's now affecting the ways they parent their children. Facilitators should be able to model a relationship that makes the participants feel safe.

Fostering dialogue between parents, therapists, social workers, and educators is essential for the development of strategies that would contribute to a conscious evolution from an egocentric culture toward a culture of solidarity, from a hierarchical society (moving in the vertical plane) to a democratic society (moving in the horizontal plane), from the adoption of norms and beliefs without questioning them, toward a mindful and self-critical culture.

It's likely a misattribution, but Albert Einstein is often quoted as saying, "Insanity is repeating identical behavior and expecting different results," and the quote has its logic. We must stop reproducing parenting styles that hurt, disconnect, exclude, and push the children to secure certain outcomes. Instead, we need to adopt more conscious and inclusive parenting methods, opening the door to learning new forms of relating based on empathy and insight.

It's challenging to act humble and vulnerable in the Western world. How can a culture of solidarity develop when the most vulnerable are precisely those who, instead of being supported, are rejected and chastened? And, what is the origin of this type of interaction? Anthropology studies suggest that it goes back to the establishment of stratified societies and that inequality is the result of the advent of agriculture. In the history of homo sapiens, power differentials between women and men, adults and children, the wealthy, and the dispossessed are a relatively new thing.

Since research shows that higher levels of empathy contribute to cooperation, academic success, and mind health, empathy training should be systematically woven into schools' curricula. But we should also be training community leaders and parents in order to promote the most successful parenting models to foster personal responsibility for our reactions and behavior and to encourage empathy and solidarity.

In the formation of compassionate individuals and citizens who are aware of their civic responsibilities, both the family, the school and the community must be involved. Many parents, professionals, and educators, disoriented about how to approach parenting nowadays, automatically perpetuate parenting systems based on punishment and reward.

## Emotion regulation and compassion training

*Being smart is a fleeting phenomenon. It is self-destructive. Why? Because intelligent life often fails to solve the biggest problem of all: the problem of cooperation.*

—Martin Nowak

Despite the fact that the anatomy of *Homo sapiens* hasn't changed much in the last two hundred thousand years, our cerebral cortex has evolved enough to allow us to consciously and voluntarily activate the tend-and-befriend response. According to Shelley Taylor, when this response is put into play, the fight-or-flight system reduces its activity, perhaps indicating that the system of caring and befriending is most effective in reducing stress.

Unconditional love toward other human beings can be learned if the conditions are favorable. From what we know about neuroplasticity, we can predict that the more we activate our circuits of empathy, the more often our brain will respond not from panic, distrust, and apprehension, but from the affiliative response. Our mind will build confidence that it can face and act on a perceived threat by leaning on support networks, forming alliances, and calling for solidarity.

Let's make it clear that no response to stress is undesirable. The three we've described play protective roles, but the mind can be trained and can contribute to regulating the emotions that mediate these responses to mitigate anxiety, fear, and the harmful effects of stress.

The disposition to help others (the compassion that motivates altruism and solidarity) is not necessarily a stable feature of one's personality. Adverse circumstances often compromise our innate inclination to experience and express compassion. Therefore, it needs to be approached as a skill that develops with age, a skill that can be and needs to be cultivated.

Helen Weng and colleagues,[155] from The Center for Healthy Minds at the University of Wisconsin, found that after only two weeks of compassion meditation training, individuals showed increased motivation toward altruistic behavior and corresponding neural changes.

Two groups listened to a recording daily. For a period of two weeks, one group listened to instructions for thirty minutes as part of a program consisting of the practice of a meditation for compassion, a technique of Buddhist origin that increases the feeling of concern toward those who suffer. So far, only monks with years of training had done this practice. In the experiment, everyone had to imagine a time when another person would be suffering. Then they focused on wishing that the suffering would go away, using statements such as, "May you be free from suffering. May you have joy and be at ease."

One group of participants began by visualizing someone close to them—a partner, family member, or friend—and directed the technique toward that person. Then they successively directed it toward themselves, a complete stranger, and finally, on a person who annoyed them, such as a coworker or a neighbor. Another group was asked to listen to a recording made by themselves that described a traumatic or highly stressful event, but they were invited to approach the event from a fresh new perspective, less charged with negative emotions, as if they were watching it from another person's point of view. This technique is known as cognitive reappraisal.

[155] H. Y. Weng, A. S. Fox, A. J. Shackman, D. E. Stodola, Z. K. Caldwell, M. C. Olson, G. M. Rogers, and R. J. Davidson, "Compassion Training Alters Altruism and Neural Responses to Suffering," *Psychology Science* Volume 24 issue 7 (1171–1180).

To assess the effects of the exercise, the subjects were asked to participate in an adaptation of the dictator-victim game, which includes sharing money, and testing altruistic behavior. Those in the first group (meditators) felt significantly more inclined to share money to help another.

Before and after the exercise, all of the participants' brains were scanned with magnetic resonance while they looked at a series of photos, some of which showed people suffering. The study showed not only an increase in altruistic behavior after compassion training but a concomitant activation in brain regions involved in social cognition and emotion regulation, including the inferior parietal cortex and the dorsolateral prefrontal cortex (DLPFC). Also, an increased connectivity between the DLPFC and the nucleus accumbens was observed.

The group that meditated also showed changes in the DLPFC, which is involved in the regulation of emotions and positive feelings in response to a reward. Although the brain changes were similar in both groups, in the second group there was no correlation between brain activity and more altruistic behavior. Seemingly, a greater regulation of emotions can help a person be more altruistic and sensitive to other people's suffering, instead of feeling overwhelmed by their feelings.

In the book *SuperCooperators: Altruism, Evolution, and Why We Need Each Other to Succeed* (written with Roger Highfield, Free Press, 2012), Harvard biology and mathematics professor Martin A. Nowak speaks about how compassion and cooperation have been the main drivers of evolution.

"Many problems that challenge us today can be traced back to a profound tension between what is good and desirable for society as a whole and what is good and desirable for an individual," Nowak says. "That conflict can be found in global problems such as climate change, pollution, resource depletion, poverty, hunger, and overpopulation."

A certain degree of compassion is essential for altruism and cooperation. Mutual support and concern for each other are essential values for survival. This statement is not in contradiction, as some believe, with the mechanism of natural selection. Compassion turns out to be a typical strength of human beings, though many mammals are also capable of displaying what could be interpreted as empathic behavior.

Two of the most popular compassion training programs in the United States are: Cultivating Compassion, developed at Stanford University, and offered to the public with instructors across the country. The other is Training in Compassion, offered at the Emory University in Atlanta. Both courses emphasize the practice of mindfulness.

The Stanford course lists the following as benefits of their training:

- Enhanced ability to feel compassion for oneself and others
- Increased calmness and ability to handle stressful situations
- Better engagement and communication in relationships
- Self-evaluation and mindfulness when feeling overwhelmed
- Increased job satisfaction and decreased sense of being overwhelmed by the job
- A more pleasurable, nurturing life at home or with family.

As mentioned before, Paul Gilbert, created compassion-focused therapy (CFT) and compassionate mind training (CMT). A pioneer in his field, Gilbert believes in our human potential for creativity, love, altruism, and compassion, but also for selfishness, revenge, and cruelty. These traits would be linked to the way our brains have evolved to solve various survival challenges. Understanding that people can be trained to have empathy and compassion and that the learning process involves the formation of new neural networks, Gilbert asserts that we can explore different ways of managing the potentials of our brains to develop specific skills that promote solidarity.

There is abundant literature documenting the fact that meditation and mindfulness practices have a positive effect on the regulation of emotions. Rimma Teper and his collaborators at the University of Toronto were interested in understanding the underlying mechanism. They suggest that the awareness of the present moment and the unbiased acceptance of our thoughts and emotions, cultivated through mindfulness, are crucial for promoting executive regulation of emotions, as this practice increases our sensitivity to affective cues. Executive function and self-regulation skills are mental processes essential for us in making plans, focusing attention, remembering instructions, and multitasking.

Renowned medicine professor Jon Kabat-Zinn defines mindfulness as "paying attention in a particular way: on purpose in the present moment, and nonjudgmentally." Kabat-Zinn is the founding and executive director of the Center for Mindfulness in Medicine, Health Care, and Society at the University of Massachusetts medical school. He is the developer of mindfulness-based stress reduction (MBSR). He explored how to create methods of stress reduction based on his studies of Zen Buddhism. Back in 1979, Kabat-Zinn recruited a group of chronically ill patients who were not responding to treatment and asked them to participate in an eight-week stress-management program. He obtained positive results: both the physical and mental health of the participants improved significantly. Today, his exercises are taught throughout the world in medical centers and hospitals, and psychotherapists trained in mindfulness techniques find it very useful in their work.

Mindfulness practice makes observers out of us, reducing our tendency to react with fear when facing stimuli that could trigger a fight-or-flight response. People suffering from depression, anxiety, or obsessive thoughts have greatly benefited from the practice of mindfulness.

Other strategies have also been useful in strengthening empathy. From imitation to learning to distance from a given social conflict, many factors influence our capacity to feel compassion toward others. The aforementioned trainings yield positive results with teenagers. Previous studies have shown that, due to normal hormonal changes, the narcissism typical of the teen years, and the emotional tumult of adolescence, young people may find problems empathizing.

Today, many studies are applied in schools to implement social and emotional learning (SEL) to help children regulate emotions and tune in to the feelings of their peers. Mindfulness and compassion training could help improve the emotional response of children who suffer from social anxiety, and they could also increase pro-social behavior.

The Hawn Foundation has developed the MindUP program and curriculum, which has already been implemented in classrooms by over a thousand educators throughout the United States, Canada, and the UK. The program is designed to "help children reduce stress and anxiety; improve concentration and academic performance; understand the brain science linking emotions, thoughts and behaviors; manage their emotions and behavior more effectively; develop greater empathy for others and the world; and learn to be optimistic and happy." MindUP reports that at least 83 percent of children who have participated in their programs show improvements in pro-social behavior.

Based on the knowledge we have today about the link between increased empathy and prosocial behavior, countries such as Great Britain, Canada, and Australia have implemented school programs aimed at lowering rates of anxiety and depression, decreasing bullying, and achieving better educational outcomes. See, for example, Australia's Kindness on Purpose program or Canadian's Roots of Empathy program.

On occasion adults try to offer a quick fix to a child in distress with expressions such as "everything will be okay." Instead, a teacher or a parent may invite a child to draw or to look at images so they can safely and metaphorically talk about feelings. Adults should also label their emotions and express how they feel in their bodies to help children recognize their own feelings and safely express them.

Learning to practice compassion, and therefore learning how to love, has effects not only on our body and mind but also on our interpersonal relationships and our ability to feel sympathy/empathy toward all sentient beings. Stress management techniques and emotional regulation training are successful in offsetting social anxiety and antisocial behaviors. Compassion training and fostering empathy have proven effective in the prevention of bullying and other forms of relationship abuse. But surely, in the long term, the greatest advantage of generalizing these programs would be to build more joyful and supportive societies where cooperation would be the norm. These societies could more effectively solve the world's problems by developing and implementing nonviolent proposals and programs.

## Tend the wound

On several occasions and in a very brief time, I've found that directing couples to do "homework" together (read a book, meditate, research a topic about an issue that concerns both, create a conflict-free space) helps to improve the relationship.

When couples participate in shared activities, it activates their dopaminergic system. There is pleasure in sharing and reward in crafting solutions together. If the therapist also demonstrates and reminds them to respond empathically, they benefit even more from therapy sessions. Empathic responses can be activated by using already designed exercises that include mindfulness, active listening (with sincere interest and curiosity), sharing feelings, showing appreciation (instead of taking the other for granted), and validating each other.

My approach also includes tuning in with the couple, very intent on understanding how open they are to expose their vulnerability and helping them to see what the strengths of their relationship are.

Besides teaching them the basics of the different stress responses, I use exercises that take into account the particular issues and concerns presented by the people sitting in front of me. I have learned from countless mistakes made in my many years of practice when I've tried to adapt interventions and apply exercises and "techniques" learned elsewhere. But I know some tools that every therapist could use to raise awareness. For example, helping them become aware of their unspoken communication by pointing out how far someone has chosen to sit from their partner or asking them to notice if their body language conveys openness while the other is talking. I note whether or not they look into each other's eyes and how often they initiate physical contact.

Many couples look for practical formulas to improve their communication, and no doubt there are some techniques that work. Paraphrasing or echoing what the other has just told us to verify that we understood, avoiding judgment and withholding advice, and having the intent to really understand do help. This is called active listening, and it's very useful in learning to empathize.

There is no doubt that, if communication improves, couples are more likely to solve conflicts. Gottman's love maps seem very useful in helping to fill in the gaps in information about the other and self, gaps that usually lead to incorrect assumptions and projections. We need to make conscious all those implicit core beliefs about life and love we acquired just from growing up in our family and find out how they influenced our love life.

I like it the most when partners look into each other's eyes. When chronic conflicts, anger or shame exist, the lovers, the spouses, stop looking at each other. Once they manage to look into each other's eyes again, there is a moment of recognition. I remember a couple that had been married for forty-three years. They had moved to another city, far from their children, who were all raising their own families. I don't remember what they were discussing, but suddenly she looked him in the eye and said, "I haven't seen you for some forty years." In the task of being parents, they had forgotten to be lovers, to be a couple. But love was still there.

Helping a couple remember their romance and the reasons why they got together also helps. Once an appropriate climate is created, it's important that they try listening to each other while being very present in the moment (this, also, is part of active listening). Usually, while another person speaks, rather than giving them our undivided attention, we're mentally rehearsing what we're going to say next. Learning to listen, without judging, without interrupting, without becoming defensive, is essential.

I ask that each person in a couple speak only in the first person and without pointing fingers; that is a ground rule. We avoid any attitude or word that might trigger the fight-or-flight or the dopaminergic responses. Sometimes they compete to prove who's better, wiser, smarter. Instead of taking responsibility for their part in the conflict, they look for a culprit. They often want to prove that they're right (and that, therefore, the other is wrong). They try to get my sympathy or plot with me, as did one husband who would call and try to convince me that his wife was crazy.

One couple I worked with had a chronic conflict not about the content of what each said but about the way they spoke to each other. She would come to him to share something, and before she could round up her idea, he would roughly ask her to lower her voice because she was talking very loudly. This annoyed her so much that she often turned her back without finishing what she wanted to say. This small but constant friction had to end. It was necessary to go from accusing each other, which triggered fight-or-flight responses, to taking responsibility for their own part in the interaction. Could she be more aware of his sensitivity to noise? Could he wait for her to finish her story before saying something about the volume of her voice?

Another couple who were trying to rekindle the relationship after a few months of separation found it very difficult to communicate without eliciting conflict. Their first impulse was to vent their resentment, to accuse, to point at the wound. My role felt like that of a soccer referee, displaying yellow or red cards every time they reacted defensively or aggressively, to caution that we needed to keep the office as a safe space. At the beginning, and although they honestly wanted to get back together, they were so upset with each other that it was hard for them to even revisit the good times they had shared or to make a list of the good qualities they'd seen in their partner in the first place. In order to activate an empathic response, it was essential that they could operate from circuits other than those of fight-or-flight. Couples often succeed in their purpose of healing their wounds. In the case of this couple, in the midst of recalling shared moments, looking into each other's eyes, coming physically closer, taking responsibility for their part in the relationship, the angry and sad emotions were stopped, and the flame was lit again. When they held hands, physical contact activated the third response.

When one truly listens to their partner expressing their feelings, the listener might come to say, "I didn't know what you're going through." Or, "I wasn't aware you felt that way." Or, "I had never realized how much this bothered you." The above statements would point at the fact the empathic response has been activated.

Often, at the beginning of the sessions, when a couple is talking about their disputes, things tend to get worse, but if the foundation of their relationship is a strong friendship, things will very likely improve.

One of the couples I worked with, no matter how bad their conflict seemed to be, never stopped taking out their favorite board game to play after sharing breakfast. That counter in the kitchen where they ate and played was their safe space. Even if they didn't speak to each other for the rest of the day, this moment helped hold them together.

We can all get out of the trap of the past and free ourselves from the habits and behaviors that limit us by reexamining our relationships with a beginner's mind, and by making a sustained effort to elevate our consciousness. Our relationships would be richer if we could finally be free from the beliefs forged by unspoken traditions, family patterns, and social tenets. We are very much the product of our social existence, true. But we don't have to be its slaves or its victims. When we awake to a mindful life, we modify limiting beliefs, change relationship and behavior patterns, and improve our relationship with ourselves, others, and the environment. In doing so, we're also changing our brain, creating new neural networks, and controlling which circuits get activated. Only when we're clear about the origin of these, our moorings, can we access the necessary freedom to love unconditionally.

Since the moment, several years ago, when I began to study emotional regulation, and specifically those three responses to stress we've been talking about, I've tried to find practical applications of the new knowledge in my practice.

I have basic training in mindfulness and somatic therapies, and I have adapted some elements of these practices to the strategies I use as a therapist. I've observed that committed patients, with practice and some dedication, happily use breathing exercises, learn to pay attention to bodily sensations, and even use visualization and relaxation practices to stay in the present moment and to sense what triggers the unwanted symptoms and responses. By staying tuned in and open to the subtle bodily changes that accompany certain affective states and that signal the need to regulate emotions, they effectively achieve executive control of them, leading to significant relieve of symptoms.

That's why I propose that we need to learn how to modulate our response to stressful situations, giving mindful precedence to the affiliative response (tend-and-befriend), which allows us to physiologically temper the intensity of the fight-or-flight response. The activation of the dopaminergic system is beneficial in cases in which the person feels their life, their status, or their relationship is threatened, because it helps to motivate proactive behaviors in search for solutions. The overuse of the dopaminergic circuit is counterproductive, though.

According to the current knowledge about the neuroplasticity of the brain, when people are training to acquire new responses, relationship patterns and perspectives, the brain changes, significantly improving their quality of life.

By practicing self-compassion and empathy, a person will create new synaptic networks or use existing neural pathways, reinforcing new habits and behaviors. In other words, by adopting new habits of thought and new behaviors, they will change their brains.

In the same way, a person could also create new synaptic connections to override the symptoms that follow a trauma (mental or physical), and the brain will successfully accommodate to new challenges and changes.

The introduction of new ways of thinking about life, and even the exploration of our childhood from a grown-up perspective, can lead to the resolution of old traumas and to greater resilience. Once the new synaptic connections are established, it's easier to reuse the same circuits again and again.

Jon Kabat-Zinn has made significant contributions to the development of stress reduction techniques at the Stress Management Clinic at the Massachusetts School of Medicine. The clinic's relaxation and mindfulness techniques have become so widespread that health professionals, including psychotherapists, frequently teach or recommend them in the management of cancer, coronary heart disease, and mental health problems.

In the West, other traditional practices brought from Asia and India, such as meditation, yoga, Reiki, and Qi Gong, even if they're not adopted within the comprehensive framework of the traditional practice, have become quite popular for contributing to significant reduction in stress levels.

## The perspective of love in therapy

At the level of the individual, many things can be done to advance on the path of love. Psychotherapy is definitely one of them.

As said before, the topic of love began to interest me long before I converted into a psychotherapist in 1990. Note that I deliberately use the verb *convert* because this profession requires the learning of a skill that I don't think was innate in me: to know how to listen. To listen with my ears but also with my heart and even with my body. By being aware of my empathic response or lack thereof and of how my body reacts to the interaction with the other, I can identify my own emotions. I also take charge of my own countertransference, a reaction on the therapist's side to a patient's transference. (Transference being the phenomenon by which a therapist, a patient, or a student redirect their unresolved feelings for childhood parental figures toward a therapist or an authority figure.) We need to be aware of the fact that there are unconscious reasons for our responses to clients, and we have to make sure that our feelings don't interfere with therapy.

In addition to getting formal training, becoming a therapist requires the willingness to sit for hours listening from the heart, not the physical heart, but with the other heart, which Chinese medicine considers the emperor of all organs; the one attributed by popular belief to be the center of affection and love; the one that feels broken whenever love is withdrawn from us, when someone betrays, abandons, or rejects us.

While listening to my patients fall to pieces, open up to show me their deepest sorrows as they rediscover themselves and face the fear of their own emotions, I've been able to cultivate the ability to step out of my *self* and feel with and for another human being. I still sometimes fail miserably. But in this precious experience, I can verify that in terms of our predicaments we're not very different from each other and that the remedies for our woes don't differ so much either. It's necessary to stop being self-centered and recognize that, even though we share a common humanity, the diversity of our backgrounds and beliefs makes us see the world from different viewpoints. Acknowledging this difference allows me to be truly compassionate. My meditation practice has allowed me to be more present with my patients and more available and open to empathic listening.

There is no single formula that can be prescribed to heal lovesickness or repair a relationship. It's true that each one of us has to carry our own burden, seek our own light, find our own way, and call for support, but it's also true that we—psychotherapists, neighbors, colleagues, friends, teachers—are all there to serve as witnesses, assistants, supporters, as long as we cultivate empathy and our natural impulse toward solidarity.

Although my original training as an art psychotherapist was psychoanalytic, and I've been trained in systemic family therapy, it was from the humanist psychotherapists Carlos Rogers and Irvin Yalom that I learned what I consider a key element of all therapy, which is the importance of being authentic with the patient—accessible, human—and of committing to embark on their journey with them.

In his book *Client Centered Therapy: Its Current Practice, Implications, and Theory* (Constable &Robinson 1973), Rogers defined empathy (in therapeutic practice) as "the listener's effort to hear the other person deeply, accurately, and non-judgmentally." He added that, "Empathy involves skillful reflective listening that clarifies and amplifies the person's own experiencing and meaning, without imposing the listener's own material."

Any credit we could claim as therapists comes, more than anything, from the patients who educate us as we listen to their unique stories and as they nurture our desire to serve while opening for us the mysteries of the human soul, or even when they leave us because, at a certain moment, they felt we could not wholeheartedly be with them where they needed us.

So far, I haven't had a single patient who hasn't shared with me some kind of heartache. It may be that they lost a loved one, their parents abandoned or abused them, they broke up with a partner, they didn't know how to connect with their children, or that they felt incapable of loving. Maybe they loved too much, felt they're not loved back, or didn't know how to love themselves. For all of them, their biggest struggle, and the associated emotions that make them seek help, originated in an event related to love or the lack of it.

In addition to interventions that seek to help a couple revive the relationship, taking it to an enhanced, more conscious, and healthier level of functioning, therapy is an opportunity to teach the language and perspective of love.

Therapists might help individuals become more self-compassionate so that they stop self-flagellating for what they consider their faults. They can also help clients gain a new understanding of what a relationship entails without guilt trips or shaming. Clients can learn how to educate their partners about their needs, instead of expecting those needs to be predicted. They can explain their background or what their thought processes are like. A better understanding leads to an increased capacity for empathy.

Understanding that they're both growing together, that they had closed a deal, and signed an unspoken contract (beyond the pretty vows they recited when they got married) helps the couple avoid unrealistic expectations about what the other's expressions of love should be.

In some cases, a pedagogical approach (in which one person patiently teaches the other what their needs are, why they feel anxious, and that they have a particular way of reacting) mobilizes empathy. Embracing their common humanity will also boost self-confidence and self-compassion. I tell my clients their partner is not a mind reader, so they need to state what they want and think explicitly.

In therapy, a client is also invited to take responsibility not for what happened to her as a child, but for the way she, as an adult, chooses to respond to and interpret what happened. It's also important to make a commitment to stop dwelling on the pain to allow for healing to happen.

Many therapists focus their interventions on helping the patient shift perspective—from sorrow to love, from the victim's position to that of a survivor, from egocentrism to empathic understanding, from the necessity to prove they're right to prioritizing saving the relationship.

One of the most beautiful aspects of family constellations, a modality we owe to German psychotherapist Bert Hellinger, is precisely his perspective of inclusion and love. Hellinger stated that since every person at a given time was part of a family, that person acquired the right and need to belong. Many symptoms of family dysfunction (and even physical symptoms) would be the result of exclusions. Examples of unconscious exclusion would be the child who died whom nobody speaks about. Or the first husband of the mother, who abandoned her and who nobody dares to name. That father who went to war and didn't return and who is remembered only in silence to avoid reviving the grief. Family constellations, as described by Hellinger himself, is a movement toward reconciliation. Through it, what was once separated is reintegrated, and this restores what he called "the basic order of the family." It encourages the perspective of love.

## Let's listen to the heart

It's only natural that when someone says something that bothers us, criticizes us, or tell us what we should do or be, we feel flustered, responding like a wounded child, at least for a moment. If we're in a relationship with that person (as a friend, brother, co-worker, colleague), and some degree of affection has developed, before reacting from our reptilian brain or our limbic system, we can shift to a perspective of an observer and watch how the situation is affecting our physical body (ask, Has my breathing pattern changed? Am I experiencing any tensions?), our mind (Are my thoughts focused, scattered, racing, defensive?), and our relationship with the other (feelings, emotions, thought content). Mindfulness would be an ideal tool to help us be in command of our response. It allows us to push the pause button and remind ourselves that there's no justification to believe that person is a threat to us.

When we're sure an "offending" person loves us, we may acknowledge that maybe they don't know how to express themselves in a loving way. The whole interaction takes on a new meaning when, instead of becoming defensive, we try to understand what the other is trying to tell us. Are they trying to protect us? Do they have unreasonable expectations about us? Do they fear we could be hurt if we're not acting with enough maturity? Do they think they need to take responsibility for us?

From that perspective, we could respond based on a higher understanding of motives, instead of reacting or getting entangled in a dispute. Because we have seen nice people lie or use kindness guided by their self-interest, because we don't want to hurt anybody's feelings, because we know contradicting someone can get us into bad conflict, we tend to avoid speaking up. But consider the cost of not being truthful, not speaking your mind, not defending your rights, or allowing others to bully you. We should have no fear of openly expressing feelings. We could say, “This is what I heard you saying.” Or, “This is how I feel when you say that.” Find a way of listening and speaking from the heart. This is how you build a language of love, without mistrust or suspicion, accepting that the other might have learned to communicate in a certain way without being aware of how their listener might feel or respond.

The key is for us to know (trust) from the bottom of our heart that the other person appreciates or loves us. If this is clear, we should have no reason to assume that they want to hurt us.

Because we respond defensively to a perceived threat, especially when we feel vulnerable, once we feel the other wants to hurt us, we react defensively and either get into an argument (fight) or emotionally withdraw (flight). But remember that the activation of our reptilian brain shuts access to the cortical areas of the brain, preventing the delivery of conscious and logical responses.

I could explain to the other:

*My first reaction to you was defensive. What I’m hearing is that you don’t like my way of doing things or that you think that if I don’t follow your advice, something bad can happen. So, let me think about what you just said.*

If translated into the language of love, I can take a different perspective and calmly deal with the situation, because I’m looking at it from the point of view of our bond (neural circuits that control the affiliative response are activated) and not from the perspective of fear, a threat of rejection, abandonment, competition, or injury (first and second responses are activated).

Dr. John Gottman explains that when a controversial issue is discussed in a healthy relationship and criticism is perceived, we can choose to take responsibility for our role in the matter that was brought up. The key is to listen from the heart, to hear the words of the other, the (perhaps premature) reaction of my own body that wants to escape from stressful moments, the background noise in my mind that’s full of warnings about potential dangers.

There is a difference, however, between defending your territory and becoming defensive. The first is justified and is your way of responding to injustice, protecting the boundaries that someone intended to trespass, or responding to an act that hurt you (verbal or physical aggression, slander, betrayal). In those cases, we need to dare speak clearly, set up boundaries, and express how inappropriate and unacceptable such behaviors felt to us. But, again, it’s spoken from the heart; it needs to express what we feel and how it affects us.

It’s not always necessary to defend the territory. We must first evaluate the situation and make choices. Ask yourself, what impact will what the other is saying or doing have on our relationship? How important is this relationship for me? If the relationship is really important, how will it be affected by my response?

We could imagine the relationship as a separate entity in need of nurturing, as we would nurture a child or a plant. Let's muse before we speak too quickly. I know certain words said by friends or loved ones will keep playing in my mind for years. So, I need to ponder what effect I want my words to have on another. Could I repair the hurt caused by what I do or say? Do I want to take care of or kill the relationship?

To tend to a relationship, we must know how to express our feelings or thoughts without finger-pointing, without accusations, without trying to make the other responsible for what we feel or do. If wanting to preserve it, muting the relationship is not an option. Gottman found in his research that stonewalling kills love.

## Experiencing awe makes us more benevolent

*I look up at the night sky, and I know that, yes, we are part of this Universe, we are in this Universe, but perhaps more important than both of those facts is that the Universe is in us. When I reflect on that fact, I look up—many people feel small, because they're small and the Universe is big, but I feel big, because my atoms came from those stars.*

—Neil deGrasse Tyson

I feel it mandatory to contribute a few paragraphs about awe, that feeling of dwelling on the beauty of life, transcending the ordinary and entering the realm of the vast unknown. Especially now that studies have found a positive correlation between awe and the development of empathy and a sense of well-being.

Awe is what a little child feels when he marvels at beauty or at anything new, at the unknown or at the skill he has just discovered. It's an emotion that can involve gratitude, much admiration, love, a bit of confusion, and even fear. In the midst of our routine lives, too frequently lived on autopilot, we tend to kill our capacity to feel awe.

I can't speak of awe without honoring the memory of late Hawaiian neuropsychologist Paul Pearsall, PhD, a loving presence, a voice I heard for the first time at the 2008 Annual Psychotherapy Networker Symposium, in Washington. He said, "There's never any closure in an awe-inspired life, only constant acceptance of the mysteries of life."

Pearsall wrote *AWE: The Delights and Dangers of Our Eleventh Emotion,* (Health Communications, 2007). He said of awe, "It's an overwhelming and life-altering blend of fright and fascination that leaves us in a state of puzzled apprehension and appreciative perplexed wonder."

In another of his books, *The Heart's Code: Tapping the Wisdom and Power of Our Heart Energy,* (Broadway Books, 1998), Pearsall spoke about "contextual cardiology." He and his colleagues affirmed, like the traditional Hawaiian doctors, that it's the heart that loves and feels. Pearsall was a member of a transplant unit, and his experiences led him to believe that the heart thinks, remembers, communicates with other hearts, helps regulate immunity, and stores information, constantly generating waves throughout the body.

It seems that Pearsall peeked into some amazing facts that science has not yet been able to deny or confirm. He contributed to create the Heart-Mind Institute at the Cleveland Clinic but I can't find evidence that this work continued any further after his death. In 2011, after having cooperated with the North Hawaii Community Hospital in a project involving the blending of conventional and alternative medicine, the Heart-Mind Institute ceased operations.

Joseph Chilton-Pearce, the author of *The Biology of Transcendence* (Park Street Press, 2002), explains that half or more of the heart is actually composed of neurons similar to those found in the brain but there is no evidence that these heart neurons are related in any way with the mind. According to Chilton-Pearce, heart and brain neurons are in constant communication. Eastern philosophies and medicines have for centuries referred to energy emitted by the heart.

Chilton-Pearce talked about the existence of an electromagnetic field (three fields, actually) coming out of the heart and traveling back to the heart in the shape of a torus. The fields have been detected with the help of a SQUID magnetometer. Some researchers still wonder if these electromagnetic fields emanated from the heart are related to consciousness. Should the reader be interested, I suggest exploring the subject of magnetocardiography.

Dacher Keltner believes that awe helps people feel part of a collective. As the director of the Greater Good Science Center (UC at Berkeley), he published *The Science of Awe* in 2018, a white paper by Summer Allen, Ph.D.

Another study, by Adam Hoffman[156], at the University of California in Irvine, suggests that the experience of awe inclines us to be more benevolent toward others and is beneficial to our body, our mind, and our relationships.

---

[156] A. Hoffman, "How Awe Makes Us Generous," *Greater Good Magazine* (August 2015).

Other studies report that people who have a good sense of awe are less impatient and feel they have more time available. One of these studies found that people with a capacity for wonder find more time to help others, value experiences more than material things, and generally feel more satisfied.[157]

Professor of behavioral psychology Paul Piff, at UC Irvine and UC Berkeley, has devoted himself to studying altruism, emotion, social classes, and social inequality. He reviewed the correlation between the times a person experienced awe and their inclination to be generous. Participants in one of his studies[158] reported that awe led them to gauge their sense of importance in relation to the greatness of the universe, with which they felt more connected. The consequence: they exhibited more ethical behavior, humility, and respect for the planet.

## Cooperation as antidote for competition

Learning to cooperate is the best antidote to competition, which at the global scale manifests as a frenzy to appropriate resources such as the Amazon, causing unthinkable devastation to this universally invaluable repository of biodiversity and natural resources.

If we don't learn to work jointly, we lose as individuals, as professionals, as members of a family, as a tribe, as a planet. When resources are scarce (and many resources such as water, oil, natural gas, phosphorus are already limited in some regions of the planet) it becomes necessary to return to our origins. We must start acting again according to community principles. Individualism raises the person's interests above those of the collectivity.

Consider that instead of competing, we could emulate. In emulation, the other is not perceived as a rival that needs to be eliminated in competition but as an inspiration or model. When we see what others are capable of, we will try to match their efforts and maybe even surpass them, not seeking a reward but to improve ourselves and our contribution to the world.

---

[157] M. Rudd and K. D. Vohs, J. Aaker. "Awe Expands People's Perception of Time, Alters Decision Making, and Enhances Well-Being," *Psychological Science* Vol 23, Issue (August 10, 2012) 1130–1136. https://doi.org/10.1177/0956797612438731.

[158] P. K. Piff, P. Dietze, M. Feinberg, D. M. Stancato, and D. Keltner. "Awe, the Small Self, and Prosocial behavior." *Journal of Personality and Social Psychology*, Vol 108(6), June 2015, 883-899

Instead of directing my energy to the effort of eliminating you so that I, my ideas, and my possessions stand out and predominate, I'll focus on giving the best of myself, looking for the common good, and not personal recognition or reward. The planet is in such a critical state that, sooner or later (hopefully sooner), we'll need to acknowledge the precedence of our common goals, and we'll come to fully understand that what's good for you is good for me.

## Remedy for our ailments: the other

Revealing the secrets of the soul can help us connect with others, but it can also make us vulnerable. That's the fear that prevents us from opening ourselves in the presence of others.

A good therapist, especially if offering a humanistic approach, transfers her capacity for empathy to therapy. Sometimes it serves the purpose of normalizing the patient's behavior or feelings when we say, "I understand; I've experienced the same as you." (This is affective empathy.) Or, "I imagine how hard this moment you're going through is for you." (This is cognitive empathy.) The practice of empathy, however, has not necessarily been a subject of study in the schools where psychotherapists are trained. There are exceptions, such as the new program at the University of Texas Medical Branch at Galveston, which is focused on promoting self-awareness, better communication, and becoming a healing presence for patients. Hopefully more programs like this will follow.

Peter Breggin,[159] an American psychiatrist who has been critical of shock treatments and psychiatric medication, discusses how sometimes our professional training prevents us from opening up when we work with patients. Among other things, the temptation to diagnose or categorize them in some way becomes an interference. He coined the term "healing presence," referring to a compassionate and self-conscious attitude during the therapeutic act, in which both empathy and intuition are used.

Dr. Irvin Yalom would agree, as this passage from one of his books, *Lying on the Couch* (HarperCollins, 1996), suggests: "What? 'Borderline patients play games'? That what you said? Ernest, you'll never be a real therapist if you think like that. That's exactly what I meant earlier when

[159] P. Breggin, "Empathic Self-Transformation and Love in Individual and Family Therapy," *The Humanistic Psychologist* Vol. 27, No.3 (1999): 267–282.

I talked about the dangers of diagnosis. There are borderlines and there are borderlines. Labels do violence to people. You can't treat the label; you have to treat the person behind the label."

Carl Rogers, who was both educator and therapist, discovered how effective it was to put oneself in the client's shoes. According to Rogers, there are three core conditions that can help a client to become a person (in other words, to achieve full mental health and self-realization). The three conditions are: empathy, congruence, and unconditional positive regard. Roger's humanistic therapy sought to help the client increase her feelings of self-worth and reduce the level of incongruence between her ideal self and her actual self.

Yalom—inspired, like Rogers, by a humanistic philosophy—found that empathy was a tool without par in therapy. In his book *The Gift of Therapy: An Open Letter to a New Generation of Therapists and Their Patients* (HarperCollins, 2002), Yalom also advocated for an authentic relationship with patients: "The establishment of an authentic relationship with patients, by its very nature, demands that we forego the power of the triumvirate of magic, mystery, and authority."

To feel that our deepest sorrows move the heart of the other has an indescribable quality found in that brief moment in which the protective barrier we've created around us crumbles and defeats the feeling of loneliness we often secretly carry.

Our whole system (or systems: social, health, family) is organized in a way that perpetuates the restrictive and problematic patterns in relationships.

This young adult, for example, is trapped in a deadly dependence on her mother. When contaminated by fear, attachments are not healthy. It angered her that her mother would not show caring when she most needed it, but at the same time, she experienced fear of losing her mother and felt she wouldn't be able to become a good mother herself. She compensated for her insecurity by creating codependent relationships, which lead to engaging in abusive interactions.

In this case, the daughter exploits the generosity and compassion of her mother, seeking her help whenever she's experiencing an emotional crisis, demanding exclusive attention, creating a drama that appeals to all of the mother's protective instincts but sometimes elicits non-parental feelings in her mom. Her mother feels a sense of duty or the shame of believing she hasn't been a good enough mother. The guilt of perhaps not having provided sufficient care, or for mistakes made that caused emotional pain to her daughter, leads the mother to play the rescuer in the drama. If this daughter becomes mentally ill, if she threatens to commit suicide, if she has to be hospitalized, the mother might become so overwhelmed that she will determine that the healthiest thing to do is to cut her daughter off because of how unhealthy their interactions have become, and because it so hurts her. Without warning, without transition, causing a tremendous wound to the dependent daughter, she disappears from her life. Mom's guilt increases and, after some time, the mom yields once more to the temptation to help the daughter and relieve her own distress.

How do you free yourself from these traps? How do you heal relationships?

As we've explained, neuroscience offers us a novel perspective. Now we know we aren't condemned to repeat what we've always done, and we have learned about the possibility of changing the brain and the way we experience emotions. Changing our viewpoint leads to ending repetitive behaviors and to a suitable regulation of our responses to stressful situations.

Even the conceptualization of mental illness has been affected by what we know today about the brain. Treatments can no longer be limited to the cognitive and the behavioral. Pharmaceuticals do relieve symptoms, yes, but they don't transform us. They can't change brain circuits. Therapy, mindfulness, exercise, and new experiences can.

We need holistic, systemic approaches to contextualize healing practices. We need to find and apply treatments that contribute to developing the areas of the brain that can make us more empathetic. Changing our perspective and our physiological responses will prove more effective than treatments focused on changing behaviors. These treatments must take into account people's context, lifestyle, and limiting beliefs.

Becoming more knowledgeable and skilled at understanding the ways we respond to stress opens a new path not only in the way we raise and educate our children, but also in the field of individual and couples' therapy. The psychologist Alan Godwin, author and publisher of *People's Problems* has developed a step-by-step technique to help others learn to recognize our predominant stress responses. Being aware of what pushes our buttons, and which is our primary response to stress is certainly useful.

But what about our second response? It would be good for us to realize that there are healthy and unhealthy ways of seeking pleasure. Sometimes we unconsciously seek pleasure in competing, trying to prove the other wrong or show that we know more. Are we willing to sacrifice a relationship to prove that we're right, that we know more? Very often we are.

What would the course of therapy be if we were able to train ourselves and our clients in compassion through meditation, visualization, the practice of mindfulness, or any other viable method? My hypothesis is that if we're equipped for love, if we train and cultivate empathy and compassion, the symptoms afflicting *Homo sapiens* at both the individual and group level will eventually be relieved, and we will ultimately save the planet.

In my understanding, the process of healing, the therapeutic process, is nothing more than a series of steps toward a greater awareness—awareness of oneself, of the body, of the secondary gains we get from symptoms, of how insane certain patterns of interaction can be. It leads to awareness of our individual and group rights, of the need to establish limits, of our capacity to be compassionate.

We grow progressively through the stages of love I proposed above in a process that includes the acceptance of self and of the other, taking of responsibilities, developing empathy, and being willing to be of service. But we have to be aware of the direction we're taking in order to determine the path we must travel.

For many people, psychotherapy is a bit of a taboo. There's a fear associated with it, and a social stigma. And some people feel there's something wrong with them if they're not able to face their struggles on their own. There's a fear of being looked at with pity or distrust, as if going to therapy would reveal an intrinsic flaw. Also, of course, there is a reluctance to discuss contentious issues, our darker areas, or family secrets with a "stranger." However, the therapist is the only one trained to offer unconditional positive regard: the acceptance of the client as they are without judgment or critical evaluations.

If not a therapist, perhaps all of us should adopt a guardian or mentor to turn to throughout our lives. Someone who could serve as a mirror and walk us through the process of learning to love so we can advance through life in the least traumatic way possible. That someone would be a person we could count on in times of crisis and who would not become overwhelmed by our grief. A tutor would help provide a sense of purpose and direction and would model for us the most empathetic and compassionate relationship possible while transmitting wisdom.

## RECOMMENDED READING

Casabianca, S. *Regaining Body Wisdom*, Naples: Eyes Wide Open, 2008.

De Waal, F, *The Age of Empathy: Nature Lessons for a Kinder Society*, New York: Three Rivers, 2009

Ekman, P. *Moving toward Global Compassion*, New York: Paul Ekman Group, 2014.

Fisher, H. *Why We Love: The Nature and Chemistry of Romantic Love*. New York: McMillan, 2004.

Gilbert, P. *The Compassionate Mind: A New Approach to Life's Challenges*, Oakland, Harbinger, 2009.

Gilbert, P. *Overcoming Depression*. London: Robinson, 2009.

Hughes, D. and Baylin, J. *Brain-Based Parenting: The Neuroscience of Caregiving for Healthy Attachment*, New York: W. W. Norton, 2012.

Keltner, D. *Born to Be Good: The Science of a Meaningful Life*. W. W. Norton & Company. Kindle Edition.

Miller, A. *Thou Shalt Not Be Aware: Society's Betrayal of the Child*, New York: Meridian Books, 1986.

McDevitt, J. B. and Calvin F. Settlage (Eds.), *Separation-Individuation: Essays in Honor of Margaret Mahler*, New York: International Universities Press, Inc., 1971.

Perel, E. *Mating in Captivity. Reconciling the Erotic + the Domestic*. New York: HarperCollins e-books, 2007.

Pinker, S. *The Better Angels of Our Nature: Why Violence Has Declined*, New York: Penguin, 2011.

Rifkin, J. *The Empathic Civilization: The Race to Global Consciousness in a World in Crisis*, London: Penguin Publishing Group. Kindle Edition, 2009.

Seppala, E. M., Simon-Thomas, E. Brown, S. L. Worline, M.C. Cameron C. D, and Doty, J.R. eds. The *Oxford Handbook of Compassion Science*. New York, NY: Oxford University Press, 2017.

Siegel, D. J. *The Mindful Brain*, New York: Norton, 2007.

Szalavitz, M. and Perry, B. D. *Born for Love: Why Empathy Is Essential—and Endangered*, New York: HarperCollins e-books, Kindle Edition, 2010.